# System Dynamics for Complex Problems in Pavement Engineering

Increasingly, segments of the civil infrastructure are considered to be parts of larger systems, which requires a systems approach for a fuller and proper understanding of and solutions to problems. Unfortunately, the subject of a system or a systems approach is barely covered in a standard civil and environmental engineering curriculum. Most, if not all, civil engineering problems involve interdependency, and hence segmented approaches of learning one individual topic at a time make it difficult for students to learn, understand, and apply rational concepts for the design, construction, and maintenance of larger infrastructure components. *System Dynamics for Complex Problems in Pavement Engineering* presents an introduction to a systems approach to help readers evolve and develop their capabilities of learning, communicating, and researching through system dynamics modeling and experimentation.

Furthermore, it helps students appreciate the need for systems thinking in modeling, analyzing, and proposing solutions for multidisciplinary problems in pavement engineering.

# System Dynamics for Complex Problems in Pavement Engineering

Rajib B. Mallick

## CRC Press
Taylor & Francis Group
Boca Raton  London  New York

CRC Press is an imprint of the
Taylor & Francis Group, an **informa** business

First edition published 2023
by CRC Press
6000 Broken Sound Parkway NW, Suite 300, Boca Raton, FL 33487-2742

and by CRC Press
4 Park Square, Milton Park, Abingdon, Oxon, OX14 4RN

*CRC Press is an imprint of Taylor & Francis Group, LLC*

ISBN: 9781032382906 (hbk)
ISBN: 9781032385563 (pbk)
ISBN: 9781003345596 (ebk)

DOI: 10.1201/9781003345596

Typeset in Times
by codeMantra

*This book is dedicated to the memory of*
*Professor Jay W. Forrester of MIT*
*who created System Dynamics*

# Contents

# Preface

The economic development of a country is often measured in terms of its civil infrastructure system. This system provides shelter, energy, and water; connects people and places; and serves as a vital link for defense. In most cases, this infrastructure is constructed with public tax money by the government, and in general, the expense of constructing and maintaining this infrastructure is huge; however, governments find it worthwhile to invest in the civil infrastructure as it leads to job creation and boosts the economy. Design, construction, and maintenance of civil infrastructure is a significant worldwide industry.

Designing, constructing, and maintaining the civil infrastructure in good and safe condition pose a significant amount of challenge because of the numerous factors involved. First, components of the infrastructure such as buildings, bridges, and roads are built in different types of terrain consisting of different soils, water tables, and climates. Second, the infrastructure affects citizens at different levels of society and the economy, and often, its breakdown leads to a greater amount of distress for disadvantaged communities. How can the civil infrastructure be equitable? Third, for economic practicality, these structures, which consume a significant amount of construction materials, need to be built with mostly locally available materials. The bulk of this material is natural mineral aggregates, which cannot be grown back for millions of years. How can civil infrastructure construction be sustainable? Finally, these structures are exposed to the weather year-round, day and night, throughout their lifetime, and hence they must be resistant to the effects of temperature and water. How to make them more resilient to be functional in different weather and climate conditions?

Therefore, in addition to conventional design challenges, the civil infrastructure system poses significant challenges in terms of equity, sustainability, and resilience. A compounding factor here is the fact that the infrastructure components are connected to a number of disciplines – economics, society, and environment. At the same time, construction uses mixes of different types of materials; hence, the civil infrastructure consists of "systems" at various levels, as well as "system of systems". Many components of these systems are interdependent and have feedback, and many of the actions have delayed, nonlinear, and/or cascading effects. Quite often, the right "leverage" is not utilized in infrastructure planning and construction to cause significant, long-term, positive changes, and therefore many "solutions" tend to have short-term benefits but long-term ill effects. Furthermore, many infrastructure problems do not have conventional analytical solutions; hence, the only way to understand them and develop solutions that are good in the long term is through modeling and simulation. Proper addressing of these challenges requires a systems approach – thinking and modeling of the problems.

This is where the application of system dynamics (SD) comes in. SD provides a methodology and tools to model problems from a systems perspective, with appropriate components and connections, and then simulate them to observe the interactions, feedback, and effects over time. SD provides a platform to logically integrate components from various disciplines in a coherent and rational manner to build a comprehensive model that can be simulated over time. SD is extremely helpful to consider the viewpoints of people and illustrate problems and simulations to them from various disciplines, and thus help them reach a consensus – this is critical for the implementation of results, specifically in terms of policy – be it for engineering specifications or economic or social changes. One of the great benefits of SD is the practice of modeling – which forces the researchers to take a comprehensive view of the problem, consider various components, be empathetic to different types of views, and work toward a solution that addresses several issues (multi-solving) at the same time.

This book is offered as a first course on applications of SD specifically for civil engineering students and researchers, with examples from pavement engineering. In addition to learning the application of SD, the content of the book is particularly relevant to civil engineers because of the emphasis on sustainable and resilient transportation all over the world; there is a great stimulus in the United States, with the Federal Highway Administration's $6.4 Billion Carbon Reduction Program, the Bipartisan Infrastructure Law, and the Infrastructure Investment and Jobs Act (IIJA). In the coming year, states and other entities will be trying to come up with a vision, climate change resilience, training, tech transfer, and concepts for centers of excellence in resilience. In this regard, this book will be significantly helpful for both students and practitioners to explore systems concepts and develop insights for sustainable and resilient transportation systems.

Chapters 1 and 2 provide the basis for this book and the pedagogical reasons, respectively. These chapters should help the teachers to decide when to offer a course in SD, to train the students and, at the same time, meet some of the critical requirements of engineering education. Chapter 3 provides examples to illustrate the dynamics in pavements. It will help students identify the dynamic nature of problems that they encounter and their importance. Chapter 4 gives an introduction to modeling – how to model a problem (not a system) from a systems perspective. This will help students prepare for modeling and simulation. Chapter 5 provides examples of different types of relationships between variables. It will help students understand the reasons behind systems behaviors that are observed in general.

Chapter 6 provides the key principles of SD, with examples from pavement engineering. In Chapter 6, students will find the principles of cascading effects and non-proportional effects in a system through an example of the behavior of multiple sections on a roadway pavement. The concept of feedback loop is illustrated here with examples. Chapter 7 further reinforces the concept of feedback loop through an example of a pavement structure system where the initial damage of one layer is shown to trigger the damage in different layers and reinforce its own damage over time. Chapter 8 provides an introduction to the analysis of systems through two examples of mixes in pavement engineering. It illustrates the application of SD to understand the behavior of mix systems as a result of changes in the behavior of the components over time. This chapter will help students understand the multidisciplinary nature of problems and consider a comprehensive view of modeling and simulation, specifically when a balancing act is required to satisfy multiple criteria.

Chapter 9 provides three examples of multidisciplinary considerations in SD modeling. It helps students consider a comprehensive view of the problem by identifying the critical factors from various disciplines that contribute to the problem and developing simulations that provide information for different factors and how to combine them into composite indices for making decisions. The examples pertain to sustainability and the impacts of climate change. Chapter 10 combines SD with probabilistic considerations – this will help students extend the application of SD to the realm of prediction from observations and understanding. It also combines the disciplines of engineering and economics for developing a rational method for investment to improve the resilience of systems. Chapter 11 will help students understand the modeling of system of systems, with an example of the resilience of a major coastal airport. It shows the resilience of systems against coastal storms, as measured by different indices, and the dependence of systems on other systems for the overall resilience of the airport.

In addition to providing guidelines on some key concepts such as ideal steps and mathematics in SD modeling, Chapter 12 is intended to further reinforce the importance of SD in students and researchers through a discussion of impactful concepts that have been developed through SD modeling, and the relation between systems science, systems engineering and SD. Some essential readings are also provided.

I hope that students, teachers, and researchers will find this book useful, and I welcome their suggestions for improvements.

# Acknowledgments

I am indebted to Professors Khalid Saeed and Michael J. Radzicki (Worcester Polytechnic Institute, WPI) for introducing me to the field of system dynamics. I am extremely grateful to my former student Mr. Frederick Kautz for developing many of the models that have been presented here. Special thanks go to Dr. Karim Chichakly (ISEE Systems, Inc.) for his help in modeling. I am thankful to the System Dynamics Society (https://systemdynamics.org/) for organizing several excellent presentations by renowned researchers and practitioners, which further helped me learn and understand system dynamics. I am also indebted to the wonderful staff of the Taylor and Francis group for making this book possible.

This book would not have been possible without the help, advice, and patience of my wife, Sumita, and daughter Urmila – I am grateful to them.

# Author

**Rajib Basu Mallick**, PhD, PE, FASCE, is Visiting Professor of Civil Engineering at the University of Texas, El Paso, and an Affiliate Professor at Worcester Polytechnic Institute in Massachusetts, USA. He is a graduate from Auburn University, USA, and has worked as a senior research associate at the National Center for Asphalt Technology (NCAT), Auburn University, as a Ralph White Family Distinguished Professor at Worcester Polytechnic Institute in Massachusetts, USA and as a Professor of Civil Engineering at IIT Delhi, India. He has more than 30 years of experience in the field of pavement engineering, specifically resilient and sustainable pavements. He has successfully completed more than 30 research projects and 10 consulting projects for the Federal Highway Administration, the Federal Aviation Administration, and State Departments of Transportation, and has developed technologies that are used for the design of innovative bituminous mixes. He is a recipient of the US Fulbright fellowship, has authored 2 textbooks, 106 journal papers, 75 conference papers, and 38 research reports, and has a US patent.

# 1 Why this Book?

The Internet and published literature are replete with the word "system". More and more parts of the civil infrastructure are being looked as parts of systems which require a systems approach for proper understanding of and solutions to problems. Unfortunately, the subject of system or systems approach is barely, if at all, touched in a standard civil and environmental engineering curriculum. How can we then expect students to be familiar with these concepts and proficient in solving tomorrow's infrastructure problems?

Civil and Environmental Engineering (CEE) consists of learning of basic principles; communication of ideas and results; research and implementation of principles; and results of research for the design, construction, and management (maintenance and rehabilitation) of the built environment. Most, if not all, problems in CEE involve interdependency, and hence segmented approaches of learning one thing at a time make it difficult for students to learn, understand, and apply rational concepts for design, construction, and maintenance of infrastructure components.

This book presents an introduction to "systems" approach that helps us evolve and develop our capabilities of learning, communicating, and researching, by "system dynamics" modeling, and experimentation through simulation of the models of pavement engineering. The basic principle of this approach is the use of "systems thinking". This approach is considered to be advantageous over conventional "segmented" learning because it helps us consider "problems" rather than individual concepts and, hence, helps us understand the interactions between all the important components that are responsible for a problem.

## WHY MODEL?

Thinking through simulation of models is the most important step for proper learning, communication, and research. It also helps us answer the question "what should I learn and why"?

We develop preliminary models on the basis of our mental models that are based on our knowledge and experience, compare the outcomes of simulations with known data or scenario or common sense, criticize the preliminary models, and then further improve them, either by including more or different components and/or by improving their representations. Although we are successful in thinking through simple models (e.g., calculating the time it will take to reach the office for a given distance, weather, and traffic conditions), human minds are limited in their extent in constructing elaborate models and then simulating them over time; this is where the use of computers comes in. Fortunately, with the help of simulation software and web-based interfaces, students and practitioners can now develop, simulate, improve, and continually learn through the use of computer-based models.

DOI: 10.1201/9781003345596-1

## SYSTEMS THINKING THROUGH SYSTEM DYNAMICS

System dynamics is the science and art of representing problems of systems through appropriate computer models and simulating them for learning about the effect of interdependencies of various components of systems and understanding the behavior of the system over time. One side benefit is prediction of the future behavior of the system – but that is possible only after a proper understanding has been achieved. Furthermore, when a proper understanding of the behavior of a system is achieved, proper decisions regarding critical controllable factors can be made such that the overall effect is beneficial in the long run. Good policies can then be formulated and implemented, based on such decisions.

The key to successful utilization of system dynamics is the ability to represent a problem just adequately so that all essential components are represented, and all non-essential components are left out. Note that what is to be included and not to be included does depend on the system, but, to a large extent, also depends on the problem that we are trying to solve.

Generally, in system dynamics models, the behaviors of systems are simulated over time – which can be nanoseconds for chemical reactions or decades for environmental pollution. Although slightly different for different software, quantities (x) that can decrease or increase in value are represented as "stocks", the rate of increase or decrease (dx/dt) is represented as "flow", and parameters that can affect flows, and in most cases, that can be controlled, are represented as "converters". The links between these components are provided through "connectors", which are essentially the equations that relate one parameter to other(s). It is emphasized here that the specific equations that are used in this book are of less importance – what is important is the nature of the equation. One can always improve the equations over time, with more of analytical and/or experimental work and, hence, improve the models.

This book provides readers with an introduction to the systems approach, illustrates the process through examples of pavement engineering, and provides simulation tools for experimenting with different models. Finally, the models themselves are provided for the readers; they can download and further improve them for their own use.

The reader is strongly advised to try out the simulations by changing different parameters, critique the outcomes, and develop recommendations for improving the models. Through the model development process, students are expected to develop a scientific approach for modeling and solving problems, viewing systems as nonlinear, dynamic, and feedback-based entities, and identifying leverage points in systems. Utilizing different parameters from different disciplines will also help them develop empathetic thinking and, hence, be more effective in collaborative efforts of problem solving.

# 2 Pedagogical Background

The main premise of this book rests on the following four points.

1. Mental models form the basis of cognitive processes that are used to make critical decisions in all facets of life, including those related to engineering design and construction and managerial/executive work.
2. Although rich in detail, mental models suffer from a wide range of drawbacks that are primarily caused by the limitations of the human brain. These drawbacks include: (a) over-simplified and unstable structures, (b) inability to accurately trace out (i.e., think through) their inherent dynamics, (c) ambiguous and ill-defined concepts, (d) biased content, (e) inability to accurately consider time delays, nonlinear relationships and feedback, and (f) sluggishness in updating in response to new information.
3. Despite their inherent deficiencies, the mental models of experts and stakeholders *are necessary* to cope with the increasing interdependence and complexity of modern society's socio-economic-technical problems.
4. Systems thinking, specifically applied via the system dynamics approach to problem solving, has been shown to be particularly effective in improving the mental models of experts and stakeholders.

The systems approach has been defined as "a framework for seeing and working with the whole, [while] focusing on interrelationships between parts rather than individual parts" (Senge, 1990). While mental models can also be enhanced with other methods, Doyle, Radzicki, and Trees (2008) have demonstrated that mental models that are strengthened by system thinking are more stable, richer in content, and devoid of the sorts of biases that can be caused by environmental factors. Generally speaking, the success of the system dynamics approach in generating improved human decision making can be attributed to the development of tools and techniques for mapping out the mental models of experts and stakeholders and then having a computer accurately trace through the associated dynamic behavior. This process leads to learning, system insight, and improved policy implementation from those who participate in the modeling process (Sterman, 1994, 2000). At a more specific level, various studies (Vennix, 1990, 1996; Cavaleri and Sterman, 1997; Huz et al., 1997) have shown that the enhanced ability to understand and control complex systems, which is generated by the system dynamics approach, comes from helping:

1. Decision makers overcome cognitive barriers.
2. Students overcome barriers inherent in classroom-type teaching/lectures.
3. Decision makers develop better and more complete, complex, and dynamic mental models.

DOI: 10.1201/9781003345596-2

A comprehensive review of literature from the *American Society of Engineering Education (ASEE) Journal of Education* shows that a number of departments across the United States and abroad have realized the critical need for exposing students to a systems approach to problem solving early on in their curriculum and have taken the necessary steps to introduce the approach through curriculum reform and/or revision of the curriculum/strategic plan and course policies. In many of these cases, system dynamics has been used as a "fundamental component of the engineering curriculum" to introduce students to a systems approach. Civil and Environmental Engineering, Environmental Science, and Environmental Policy (CEE-ES-EP) are areas that are inherently multidisciplinary and exceptionally complex in nature due to an ever-increasing population with continually changing demographics, coupled with the rapid expansion of infrastructure in some regions of the world and aging infrastructure in others, a looming shortage of energy, alarming levels of greenhouse gas emissions, and the critical depletion of natural resources such as mineral aggregates.

There are many important questions that need to be answered to make sure that humanity keeps moving toward a sustainable future, and there are many time-dependent interacting elements that need to be considered to answer these questions. The only way to tackle these elements is through a dynamic system approach. In their Department Level Reform (DLR) project at the University of Vermont, Lathem et al. (2011) identified the systems thinking approach as the primary tool that enabled students to think about the social, ethical, and moral consequences of their work when analyzing and/or designing CE systems – a concept that merged well with the other goal of the university's DLR reform, that of service learning.

The author of this book intends to bring in another, more important, benefit that is expected from systems learning – the ability of students to map out their mental models, simulate them, and improve them. This will help students generate enhanced learning about complex systems by giving them the tools for assimilating fundamental concepts from different courses in their curriculum. System dynamics is precisely the tool that is needed for improving and enhancing mental models and subsequently improving the capacity to develop rich mental models and improve them over time to make them more complete, consistent, and dynamic.

Hence, the integration of system dynamics with CEE-ES-EP, with an objective of developing the capacity to construct accurate, complete, and dynamic mental models, is proposed as a *new and innovative step* in the direction of improving student learning and enhanced decision-making capabilities.

## OBJECTIVE AND APPROACH

The *overall objective* of this book is to provide materials and assessment methods for the presentation of a system dynamics course for civil and environmental engineering, science, and policy students, specifically to improve their mental models through the introduction of a systems approach during their senior/early graduate year, either as a standalone course or as modules in existing courses. The author has followed the guidelines (Demetry, 2014, Fink, 2005) in developing the materials through the critical steps of identifying situational factors, identifying learning goals, developing assessment, developing teaching and learning activities, and finally integrating all the steps.

The *specific objectives* of this book are as follows:

1. To introduce the principles of system dynamics and demonstrate them with respect to problems in pavement engineering
2. To provide students with concepts to critically analyze interrelated and feedback-dependent problems in pavement engineering
3. To help appreciate the need for systems thinking in modeling, analyzing, and proposing solutions for multidisciplinary problems in pavement engineering
4. To help develop a strategic view of pavement-related problems and solutions such that policies that would lead to benefits in the long term could be developed

It is expected that instructors will also use the materials in their respective courses in small modules; this will also help the students retain the concepts more effectively throughout their curriculum. As Johnson et al. (1995) point out, "It is important that the examples and motivation for system dynamics and control be given for systems and processes with which the students are familiar". For example, the environmental impact of economic development and climate change can be used as two over-arching themes, and various modules such as energy consumption, cost, and $CO_2$ emissions can be constructed from them. Other possible topics include water resource management (drought, flooding, rising seas) and urbanization impacts on the environment such as increasing temperatures as a result of more paved surfaces. Specifically, examples can be developed for complex (and compounding) cases, such as those related to urbanization and climate. For example, drought in California (in part due to climate change) combined with urbanization/population growth, increased risk and intensity of forest fires (in part due to climate change and also water supply problems), and declining support of hydropower/dams (and their reservoirs) lead to severe water and related economic problems.

## SUGGESTIONS FOR ASSESSMENT

Within groups, a pre-/post-design that assesses attainment of learning outcomes and changes in student attitude, interest, and belief should be used. The following approach (Doyle et al., 2008) can be adopted to measure student learning.

1. Just prior to the start of the course, a pre-test survey should be given to the students in which a system dynamics "reference mode" related to climate change (or any other suitable topic) is presented and a question is asked: What do you think caused the pattern depicted in the reference mode? The survey will guide the students through the process of "telling the story" represented by the reference mode data, including explaining important causal events, factors and variables, and the relationships among them. The students will decide how much or little information to include in their

responses and will be asked to express the degree of confidence they have in their explanations.

2. At the conclusion of the course, the students' mental model should again be assessed using the same survey instrument and procedure employed during the pre-test.

3. To avoid errors associated with asking students to express their ideas in new and unfamiliar ways, their narratives should be coded into a format of a causal loop (feedback loop) diagram by an assessor blind to the experimental condition using established techniques for identifying explicit and implicit structures in narrative text. By identifying the number of students who include a variable and counting the number of variables, connections between variables, and feedback relationships, pre-post differences in the content, structure, complexity, and dynamics of their mental models can be quantified. The extent to which mental models are shared by students will be quantified (by appropriate method to be decided), which will allow the assessment of the effectiveness of the course in building group consensus.

## EXPECTED OUTCOMES

The most important expected benefit will be changing the perspective of CEE-ES-EP students from reductionist and static to holistic and dynamic. In the short term, students will learn about a relevant concept and a tool and then apply them in their studies of other courses, while in the long run, students will apply this information to problems in their specific area of work. Pre-/post-testing of the students' mental models, along with statistical analysis of the results, could be used to assess these benefits. Specifically, during pre-/post-tests, instructors can evaluate students' understanding of an environmental problem in areas such as accuracy, completeness, identification of feedback loops, and specification of the system dynamics.

It is expected that the use of this book will help curriculums to provide an effective, useful, time-efficient, and ubiquitous way of meeting three Accreditation requirements:

1. An ability to develop and conduct appropriate experimentation, analyze and interpret data, and use engineering judgment to draw conclusions
2. An ability to apply engineering design to produce solutions that meet specified needs with consideration of public health, safety, and welfare, as well as global, cultural, social, environmental, and economic factors
3. A recognition of the need for and an ability to engage in life-long learning

## REFERENCES

Cavaleri, S. and Sterman, J. D. (1997). Towards evaluation of systems thinking interventions: A case study. *System Dynamics Review*, 13(2), 171–186.

Demetry, C. (2014). Creating Significant Learning Experience Workshop, 2014, WPI.

Doyle, J. K., Radzicki, M. J., and Trees, W. S. (2008). Measuring change in mental models of complex systems. In *Complex Decision Making: Theory and Practice* (H. Qudrat-Ullah, J. Michael Spector, and P. I. Davidsen, eds.), Berlin: Springer-Verlag, pp. 269–294.

Fink, L. D. (2005). *A Self-Directed Guide to Designing Courses for Significant Learning*, http://www.deefinkandassociates.com/GuidetoCourseDesignAug05.pdf, accessed 1/23/14.

Huz, S., Andersen, D. F., Richardson, G. P., and Boothroyd, R. (1997). A framework for evaluating systems thinking interventions: An experimental approach to mental health systems change. *System Dynamics Review*, 13(2), 149–169.

Johnson, S. H., Luyben, W. L., and Talhelm, D. L. (1995). Undergraduate interdisciplinary controls laboratory. *ASEE Journal of Engineering Education*, 84(2), 133–136.

Lathem, S. A., Neumann, M. D., and Hayden, N. (2011). The socially responsible engineer: Assessing student attitudes of roles and responsibilities. *ASEE Journal of Engineering Education*, 100(3), 444–474.

Senge, P. M. (1990). *The Fifth Discipline: The Art and Practice of the Learning Organization*, New York: Doubleday.

Sterman, J. D. (1994). Learning in and about complex systems. *System Dynamics Review*, 10(2/3), 291–330.

Sterman, J. D. (2000). *Business Dynamics: Systems Thinking and Modeling for a Complex World*, Boston: McGraw-Hill.

Vennix, J. A. C. (1990). *Mental Models and Computer Models: Design and Evaluation of a Computer-Based Learning Environment for Policy-Making*. Den Haag: CIP-Gegevens KoninklijkeBibliotheek.

Vennix, J. A. M. (1996). *Group Model Building: Facilitating Team Learning Using System Dynamics*. New York: Wiley.

# 3 Dynamics in Pavements

Pavements are structurally and material-designed engineered structures that provide smooth and safe traveling surface for traffic under different conditions of the environment. They can be either flexible, consisting of full-depth asphalt-aggregate mix (generally Hot Mix Asphalt, HMA) or HMA over granular layers, or rigid, consisting of Portland cement concrete (PCC) over granular layers (Figure 3.1). Pavements can also be constructed with HMA over PCC (composite pavements), thin PCC layers over HMA, or cold mixes (such as emulsion based) over granular layers (generally for low-volume roads). Most of the components of pavements, which typically consist of ~95% mineral aggregates come from natural aggregate resources – gravel pits and rock quarries (local resources). Most of the pavements in the world (about 95% in the United States) are flexible pavements. The binder material, either asphalt or cement, is manufactured industrially; its production involves energy and emissions. Pavements are important for our everyday lives, and they represent a significant investment from governments and state departments of transportation for society.

Pavements provide a smooth surface for driving by spreading the load through different layers (from top to bottom, surface, binder, base, subgrade) in such a way that the resulting stress in each layer is low enough to prevent excessive deformation from the vehicle loads under the regime of prevailing environmental conditions (Figure 3.2).

One important point to note is that water entering a pavement from rain or snow can cause serious detriment to the layer properties and reduce the load-carrying capacity. As a result, the same stress that would not have caused major deformation under dry conditions would cause (among other causes related to materials, mix design, construction, and overloading) excessive damage under wet conditions and, hence, lead to rapid failure of the pavement, such as in the form of cracks or excessive permanent deformation (rutting) or pumping (in rigid pavements) (Figure 3.3).

Pavements can be visualized as a system in many ways, such as those shown in Figure 3.4. The first (top) one shows a system consisting of the traffic, environment,

**FIGURE 3.1**   Flexible and rigid pavements.

DOI: 10.1201/9781003345596-3                                                                **9**

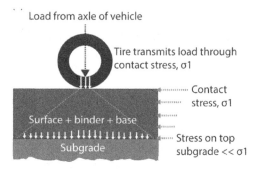

**FIGURE 3.2**   Mechanism of pavement function.

**FIGURE 3.3**   Examples of failures in flexible and rigid pavements.

and pavement. This system will be considered if the problem that needs to be solved involves the effects of both the traffic and the environment on the behavior or the performance of the pavement. The second one (middle) is a system that consists of different layers of the system. This system will be modeled if the problem involves the interactions between the layers only – for example, in a construction problem, where the condition of the underlying layers dictates the suitability of a specific type of surface layer. The third (bottom) is a system consisting of several mix components, which becomes applicable if one has a problem of modeling the interaction between different volumetric components of the mix.

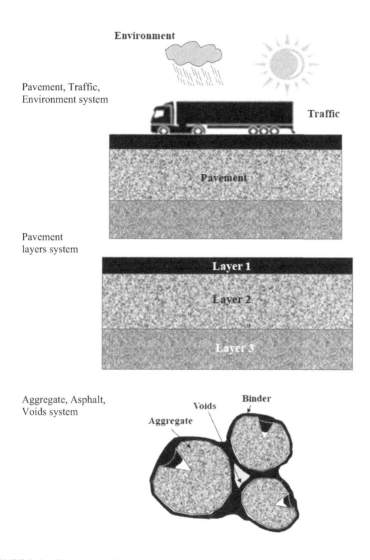

**FIGURE 3.4** Examples of systems in pavements.

Dynamics means how the various parts of the pavement system interact because of changes in the properties, and the performance of the pavement changes over time as the system changes continuously. For example, see the different properties/performances and conditions changing over time in Figure 3.5. You will notice that the changes are either monotonic or fluctuating; for monotonic, the curves may be S-shaped (fatigue curve) or concave or convex, or a "goal-seeking" curve (see the viscosity curve). For fluctuating curves, the behavior is affected by environmental factors and geographic locations. Therefore, the plots illustrate a wide range of dynamic behaviors that are dependent on various factors and represent a range of problems, which will need considerations of different types of systems for their analysis and simulation.

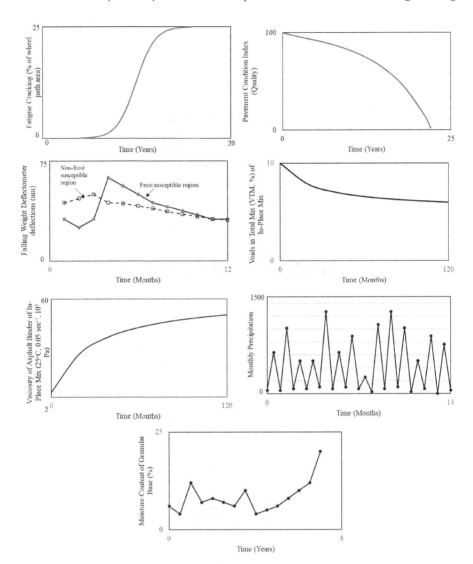

**FIGURE 3.5**   Changes in pavement properties and performance over time.

There are three important things to note in a dynamic system:

1. The "structure" or make-up of the system dictates the behavior of the system.
2. Factors interact with each other to produce a combined effect on the system.
3. There is/are feedback loop/s within the system that dictates the progression direction of properties and behavior of the system.

It is precisely with respect to the above three points that system dynamics is applicable in payment systems, as it allows us to consider the individual factors, their interactions, and feedback loops and model a relevant system – relevant to the problem at

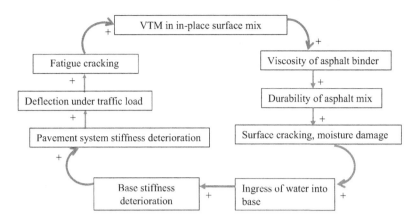

**FIGURE 3.6** Feedback loop in the performance of an asphalt pavement with a granular base layer.

hand and then predict the behavior of the system over time. The primary benefit of this approach is that the results allow us to make recommendations about altering the "structure" of the system such that its behavior over time is desirable.

An example of the detailed interaction and feedback loop is shown in Figure 3.6. This figure shows that in a pavement structure with high in-place air voids or voids in total mix (VTM), there is a potential for rapid aging of asphalt binder and a resultant increase in viscosity or stiffness. A high stiffness leads to durability problems in the form of cracking on the surface. Surface cracking/opening up of the surface leads to the ingress of water in areas from precipitation. This water enters the underlying base layer, which, in a typical flexible pavement, is made out of granular materials or aggregate. A higher saturation (caused by a higher water content) causes a deterioration of the stiffness of the base layer and the overall structural capacity of the pavement. As a result, the deflection of the entire pavement structure under traffic loading increases over time, which causes higher strains at the bottom of the asphalt mix layer, which leads to a more rapid onset and progression of fatigue cracking in the asphalt layer. With the progression of bottom-up fatigue cracking, there is a formation of more voids/opening up in the surface asphalt mix layer, which again leads to the ingress of more water from precipitation and leads to the conditions mentioned above. In essence, this becomes a self-reinforcing loop.

Note the + sign at each level indicates that each preceding parameter increases the potential of the following parameter. For this example, at some point of high damage, the concept of VTM in the surface mix will not hold – there will be potholes formed by intersecting fatigue cracks. Note that in the above interactions, we have not considered factors such as climate or traffic specifically. For example, high temperature with adequate traffic can lower the voids, and low temperature with low traffic can keep the voids at a fairly high level. However, the list of factors is sufficient to convince that there are numerous factors that are changing and also that they affect each other.

Another relevant scenario is the interaction of spatially adjacent pavement sections (see Figure 3.7). If water enters through a central line/longitudinal joint into one or two adjacent sections, then the damage caused in one section will affect the other one

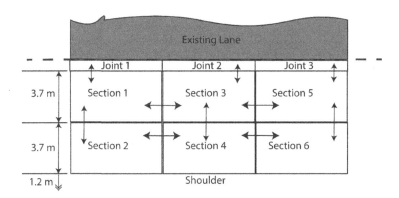

**FIGURE 3.7**   Feedback loop effects in the performance of pavements in adjacent sections.

and the damage in the second section will affect the damage in the first section resulting in a reinforcing feedback loop. The consequence is that the two sections deteriorate at a much faster rate than they would if they were physically separated from each other, or some barrier exists between the two sections that would prevent water to flow from one to the section dash; obviously, this is not the case in the real world.

The dynamics in properties and behavior of pavements require an approach for modeling pavement behavior that can capture the essential structure of the system and map the essential factors that dictate the system, with the mathematical relationships between them, such that the changes in the behavior of the system over time can be estimated from a simulation of its model.

# 4 How to Model a Problem

Most problems consist of systems with many interrelated and dynamic (change over time) factors and feedback. A proper understanding of such systems is critical for engineers who are involved in designing, constructing, and maintaining the infrastructure facilities and for senior-level personnel who are in charge of formulating policies. Such understanding is needed for knowing the behavior of systems and also for taking appropriate steps and formulating policies that lead to solution of problems and improvement of the system, namely through the alteration of its structure.

Because of the involvement of many correlated factors and dynamic nature, policies and actions related to complex systems tend to have unintended consequences, counterintuitive behavior, side effects, positive and negative feedback, nonlinear consequences, and dynamic complexity. Furthermore, behavior of such systems does not have any exact analytical solutions, which makes it difficult, if not impossible, to use theoretical calculations; the only option is modeling and simulation. Unfortunately, the human mind is incapable of simulating anything but the simplest system.

Computer-assisted modeling and simulation help us to unravel the complexities of problems, understand them, experiment with different factors, and solve them. Iterative modeling also helps us to understand a system and develop better models. Such modeling provides a "very low-cost" laboratory where one can learn through experimental simulations. At the same time, once the initial model is set up on the basis of existing information, further data collected from laboratories and the field could be used to enhance them. Therefore, computer-based system modeling should be considered as an integral part of experimental work in solving complex problems.

Note that the key word here is the "problem" – even though we refer to "systems" now and then in this book, we emphasize that the intent is to learn how to model the "problem" and not the "system". That is, we need to model only that part of the system that is relevant to the problem and nothing more.

## MODELING AS A TOOL

To model a problem, let us consider advancing through the following steps:

1. Define the problem and the boundary of the system
    a. What precisely is the problem?
    b. What are the factors, interrelationships, and feedback that are known to be associated with the problem, which will constitute the modeled "system"?
    c. What is the appropriate time scale of the problem?
    d. What are the known behaviors of the relevant dynamic factors (from existing data/literature)?
2. Develop a dynamic hypothesis of the problem
    a. Tie in the known causes and theories related to the problem

DOI: 10.1201/9781003345596-4

    b. Develop a causal structure that explains the behavior of the system in terms of the relevant factors, their changes over time, and feedback

    c. Draw the causal structure; map the different factors (parameters) with the connections between them

3. Formulate the model

    a. Assign appropriate equations and values

4. Simulate the model

5. Evaluate the model

    a. Compare the output from simulation to available information and data – Does the model reproduce the behavior of the system as evident from data from literature? (Compare the "trends" more than the "values")

    b. Is the model sensitive to changes in the key parameters?

    c. Is the model robust – behaves realistically under extreme conditions?

6. Build confidence in the model – Improve the model as needed

7. Evaluate the effect of different changes in different factors or "scenarios"

    a. What if one factor or many factors, and/or interactions change?

    b. Are the effects of the different factors synergistic or compensatory?

Example 1. Consider a very relevant problem of modern times. We are all concerned about the emission of greenhouse gases, specifically $CO_2$, from burning of fossil fuels. In this respect, we are curious to know what would be the level of $CO_2$ in the atmosphere by the end of the century, if the emissions go unabated. Here is a list of the specific details:

1. The problem is modeling an appropriate system that will be able to show us the trend in the level of $CO_2$ in the atmosphere for the next 100 years.

2. The known factors that affect atmospheric $CO_2$ levels are as follows: current $CO_2$ level (Gigaton, Carbon, GTC), rate of emission of $CO_2$ (per year), the rate of removal of atmospheric of $CO_2$ (per year) by the oceans, and the rate of removal of atmospheric $CO_2$ (per year) by land.

3. From measurements, we have the following data at hand:

    a. In 1997, atmospheric $CO_2$ level = 780 GTC

    b. Observed emission rate of $CO_2$ (from burning of fossil fuel):

| Year | Rate (GTC/year) |
| --- | --- |
| 1997 | 6.6 |
| 1998 | 6.6 |
| 1999 | 6.6 |
| 2000 | 6.8 |
| 2001 | 6.9 |
| 2002 | 7.0 |
| 2003 | 7.4 |
| 2004 | 7.8 |
| 2005 | 8.1 |
| 2006 | 8.4 |

4. Rate of removal of atmospheric $CO_2$ by oceans = 2 GTC/year.
5. Rate of removal of atmospheric $CO_2$ by land = 1 GTC/year.

It is interesting to note that the rate of emission (emission per year) is not constant but is varying with time nonlinearly; how can we model this? One hypothesis for explaining the nonlinearity is that the more fossil fuel we burn, the more opportunities we create for burning fossil fuel. That is, the amount of $CO_2$ from fossil fuel burnt at any time is affecting the rate of generation of $CO_2$ from fossil fuel by creating more mechanisms of burning fuel. Symbolically, this is shown in Figure 4.1.

The overall problem can be symbolically mapped as in Figure 4.2.

Let us gently introduce ourselves to the language of system dynamics now. In system dynamics language, when we draw a map of the problem, anything that accumulates or decreases is known as a **stock** and generally denoted by a rectangle, whereas the rate is denoted by a valve and is known as the **flow**. Therefore, the above problem can be represented as in Figure 4.3. Note that those flows that result in an increase in the stock are known as **inflows**, whereas those that result in a decrease are known as **outflows**. The arrows (red color) are the **connectors** that indicate the connections between the stocks and the flows. For example, see the top connector between "$CO_2$ generated by burning fossil fuel $CO_2$," (Stock) and "$CO_2$ generated per year" (flow) – as shown in Figure 4.1.

In this model, since 780 GT was the initial atmospheric $CO_2$, and 90% of the emissions are known to be from fossil fuel, we can assume that the initial $CO_2$

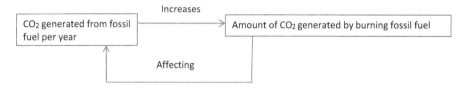

**FIGURE 4.1** Model relating $CO_2$ and generation rate.

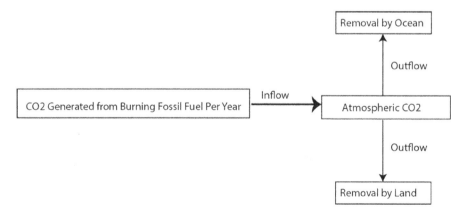

**FIGURE 4.2** Updated model relating $CO_2$ and generation rate.

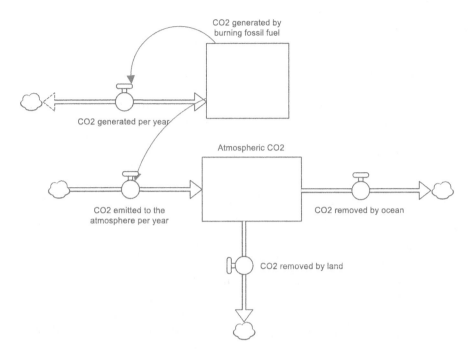

**FIGURE 4.3**   System dynamics representation of the $CO_2$ generation model.

generated from burning fossil fuel is (0.9*780) GT. Since we have decided that the rate is affected by $CO_2$, we can indicate the rate as a multiplier time the $CO_2$ generated by burning fossil fuel. For the time being, let us consider the multiplier to be 0.009 per year. Remember, this number can be updated on the basis of information/ data. The parameters and equations can be written as given in Table 4.1.

The simulation of this model involves calculating the $CO_2$ generated every year by burning fossil fuel, the rate of generation, the $CO_2$ removed by land and ocean, and the amount in the atmosphere, for every year, up to, say, 100 years. The model is simulated by numerically integrating the rate of change of stock (flow) over a time period, and this can be achieved easily by using appropriate software, such as object-oriented software, without having to write any code from a scratch (Example of software: Stella from ISEE Systems, Vensim from Ventana Systems). Note that this model was created using STELLA v 10.0.4 (iseesystems.com); you can find help on using this software to create models at:

http://www.iseesystems.com/community/downloads/tutorials/ModelBuilding. aspx

Note that throughout the book we have used STELLA, but the reader can use any other software of choice (such as Vensim [http://vensim.com/], DYNAMO [https:// github.com/bfix/dynamo], iThink [isee systems], and Powersim Studio [https://powersim.com/]). In most cases, software companies allow free, scaled-down versions of the software for use by students. Later, we will discuss the use of MS Excel for constructing a simple model also.

## TABLE 4.1
## Equations

Top-Level Model:

Atmospheric_$CO_2$(t) = Atmospheric_$CO_2$(t–dt) + ($CO_2$_emitted_to_the_atmosphere_per_year - $CO_2$_removed_by_ocean - $CO_2$_removed_by_land) * dt

INIT Atmospheric_$CO_2$ = 780

INFLOWS:

$CO_2$_emitted_to_the_atmosphere_per_year = 0.9*$CO_2$_generated_by_burning_fossil_fuel

OUTFLOWS:

$CO_2$_removed_by_ocean = 2

$CO_2$_removed_by_land = 1

$CO_2$_generated_by_burning_fossil_fuel(t) = $CO_2$_generated_by_burning_fossil_fuel(t – dt) + ($CO_2$_generated_per_year) * dt

INIT $CO_2$_generated_by_burning_fossil_fuel = 6.6

INFLOWS:

$CO_2$_generated_per_year = 0.0095*$CO_2$_generated_by_burning_fossil_fuel

{ The model has 6 (6) variables (array expansion in parens).

In root model and 0 additional modules with 0 sectors.

Stocks: 2 (2) Flows: 4 (4) Converters: 0 (0)

Constants: 2 (2) Equations: 2 (2) Graphicals: 0 (0)

}

---

The results from the simulation are shown in Table 4.2 and Figure 4.4. It can be seen that the predicted trend of emission matches closely with the trend of given data, and that is an indication of the validity of the model. Inherently, when we have considered the $CO_2$ emission rate (per year) as a function of the $CO_2$ due to emission, we have assumed a function as $S = S_0 e^{gt}$, where S is the $CO_2$ due to emission, $S_0$ is the initial value (0.9*780), g is the growth factor (0.009), and t is the time. This is an example of exponential growth, and it can be seen that according to this prediction, the amount of $CO_2$ in the atmosphere will be more than double by the end of this century. It shows the cumulative effect of generated $CO_2$ on atmospheric $CO_2$.

Therefore, we have successfully modeled the problem and obtained some useful information. If the trend of emissions did not match with the observed trend, we could have either changed the form of the relationship between the variables and introduced other relevant variables or changed the numeric value of the variable(s).

When you run the simulation of your model, you try to evaluate the model by comparing the results of the simulation with the known information or data. That is, if you use information/data in the model, for which the observations are already known, and the results of the simulation match your observations (which can be from your personal experience or the literature or experimentation) then you know that the model is a good one (remember, the model can always be improved, and there is no perfect model), or at least, a "workable" model. This mode of simulation that compares the data against already available data is known as running the "**Reference**" mode. For comparison of data from simulation in the Reference mode, try to plot the

**TABLE 4.2**

**Results of Simulation from the Model Relating $CO_2$ and Generation Rate**

| Year | $CO_2$ Generated by Burning Fossil Fuel, GT | Atmospheric $CO_2$, GT |
|---|---|---|
| 0 | 6.6 | 780.0 |
| 1 | 6.7 | 783.4 |
| 2 | 6.7 | 786.9 |
| 3 | 6.8 | 790.5 |
| 4 | 6.9 | 794.1 |
| 5 | 6.9 | 797.7 |
| 6 | 7.0 | 801.5 |
| 7 | 7.1 | 805.3 |
| 8 | 7.1 | 809.1 |
| 9 | 7.2 | 813.1 |
| 10 | 7.3 | 817.1 |
| 20 | 8.0 | 860.8 |
| 30 | 8.8 | 911.9 |
| 40 | 9.6 | 971.1 |
| 50 | 10.6 | 1,039.1 |
| 60 | 11.7 | 1,116.9 |
| 70 | 12.8 | 1,205.5 |
| 80 | 14.1 | 1,305.8 |
| 90 | 15.5 | 1,419.1 |
| 100 | 17.0 | 1,546.6 |

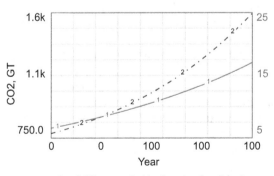

$-^1-$ CO2 generated by burning fossil fuel

$-^2-$ Atmospheric CO2

**FIGURE 4.4**   Pots of results of simulation of the model.

variables of the same unit on the same axes, and do not omit the graph of any variable if quantitative data for that variable are not available and only qualitative data (trend) are available. Remember, when modeling, the trend of the results is more important than the actual values.

In the next chapter, we will take a look at the rationale behind the use of different types of relationships (such as exponential, which was used between $CO_2$ in the atmosphere and the rate of generation of $CO_2$ in the above example).

# 5 Relationships between Variables

In the last example, we noticed that we were successful in replicating the nonlinear trend in the rate of emission (emission per year) by considering the fact that the rate of emission is a function of the current emission.

How did we decide to do this? It all starts with a (*a* model, not *the* model – note that one can very well disagree with this model and have a separate model) mental model. The model can be stated as follows: "Burning of fossil fuel (to manufacture, 'develop' etc.) creates more opportunities for burning fossil fuel". For example, one group of policy makers are of the opinion that building more roads by burning fossil fuel leads to the use of more personal vehicles or fossil fuel–driven machines, and hence leads to burning of more fossil fuel.

There is, hence, a reinforcing effect of burning fossil fuel – it leads to burning fossil fuel at a higher rate. Pictorially, this can be expressed as follows. The + signs indicate that the emitted $CO_2$ level is positively affecting the rate of emission of $CO_2$, and the rate of emission of $CO_2$ is positively affecting (increasing) the emitted $CO_2$. The R (Figure 5.1) with the circular arrow indicates that this is a reinforcing "loop" or what is known as **Feedback** loop in system dynamics.

In this figure, we indicate the following:

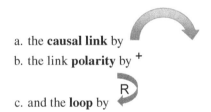

a. the **causal link** by
b. the link **polarity** by $^+$

c. and the **loop** by

From this mental model, we can write down a mathematical model, as follows:

$$dS/dt = f(S)$$

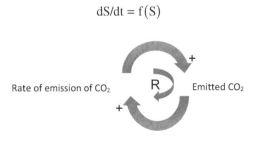

Rate of emission of $CO_2$      R      Emitted $CO_2$

**FIGURE 5.1** Feedback loop: reinforcing loop.

DOI: 10.1201/9781003345596-5

where, $S$ = value of stock (emitted $CO_2$)
and $dS/dt$ = rate of emission of $CO_2$

If $g$ is the fractional growth rate (unit of $g$ is 1/time, where time can be in years), we can also write

$$dS/dt = gS$$

Integrating this equation leads to the following equation:

$$S = S_0 e^{gt}$$

where $S$ is the stock at any time ($t$), and $S_0$ is the initial stock.

When we model this in system dynamics (we used Stella), we indicate this model as in Figure 5.2.

We assign a value of the flow as 0.009/year ($g$) and the initial value of the stock as 0.9*780 ($S_0$). We then create another section of the model where we link the above part with the part that indicates the accumulation in the atmosphere and the removal by land and ocean (See Chapter 4). Note that when we simulate the model, the software integrates the flow over time and accumulates the results in the stock. Hence, we need to specify the rate of increase (or decrease) and the initial value of the stock, to begin with. Also note that although in this case the flow is unidirectional, there can be flows that are bidirectional, that is, the flows can be positive and negative (increasing and decreasing).

The graph of generated $CO_2$ against time will be of the form shown in Figure 5.3: Note that this graph is drawn from the data that were obtained in the last example, as indicated in the last column of Table 4.2. We call this model a **first-order** (there is only one stock in the system) **positive** (reinforcing loop) **feedback system**.

Now, suppose, we run into a problem where the stock decreases over time, and the rate of decrease is affected positively by the amount of stock at any time. That is, the state can be represented as in Figure 5.4. Note that the polarity of the link from the stock to the flow is still +, but the polarity from the flow to the stock is −.

In system dynamics, this can be represented as shown in Figure 5.5.

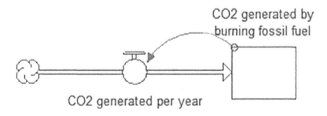

**FIGURE 5.2**  Model 1 of reinforcing loop.

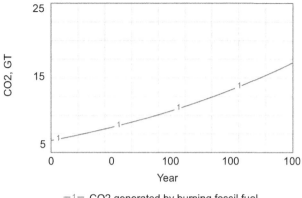

FIGURE 5.3  Plot of $CO_2$ versus time.

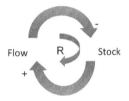

FIGURE 5.4  Feedback loop: reinforcing loop.

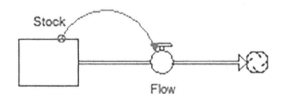

FIGURE 5.5  Outflow from stock.

## MODELING A PROBLEM WITH STOCKS, FLOWS, AND CONVERTERS

Components of the infrastructure such as roads and bridges deteriorate over time and need rehabilitation at some point. A problem that can be modeled shows deterioration over time and tells us how many bridges, say, will be ready for rehabilitation. Consider the data that are available for a state or province:

1. Initial number of bridges = 40
2. Number of bridges built every year = 1
3. Average life of a bridge = 30 years
4. Initial number of bridges requiring rehabilitation = 0

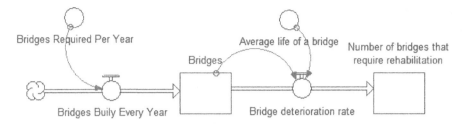

**FIGURE 5.6**   Bridge model.

The map of the problem is represented in Figure 5.6.
Mathematically, we can write the following equations.

1. Bridges required per year (converter) = 1 bridge
2. Bridges built every year (flow) = bridges required per year
2. (Initial) bridges (stock) = 40 bridges
3. Average life of a bridge (converter) = 30 years
4. Bridge deterioration rate (flow) = bridges*(1/average life of a bridge) bridges per year
5. (Initial) number of bridges that require rehabilitation (stock) = 0

The objective is to determine how many bridges will require rehabilitation over the time span of 50 years.

To simulate the model over a time span of 50 years (at a time step of 1 year), we use STELLA 10.0.4 and end up with the results in Table 5.1.

The model has been set up with two **converters** – the number of bridges required (and built) every year and the average life of the bridge. These two parameters can be changed, and the effect of the change(s) on the number of bridges that require rehabilitation can be evaluated.

Can you evaluate the effect of requiring five bridges every year and having an average bridge life of 50 years? Can you also include a converter to consider a factor that limits the number of bridges built every year?

**TABLE 5.1**
**Results from the Bridge Model**

| Year | Bridges That Require Rehabilitation |
|------|-------------------------------------|
| 0    | 0                                   |
| 10   | 13                                  |
| 20   | 25                                  |
| 30   | 36                                  |
| 40   | 47                                  |
| 50   | 58                                  |

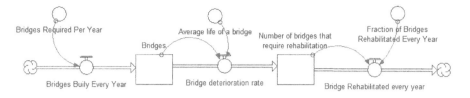

FIGURE 5.7   Revised bridge model.

TABLE 5.2
Results from Modified Bridge Model

| Year | Bridges That Require Rehabilitation |
|------|-------------------------------------|
| 0 | 4 |
| 10 | 4 |
| 20 | 4 |
| 30 | 4 |
| 40 | 4 |
| 50 | 4 |

Next, consider that the bridges are rehabilitated at a specific rate (bridges rehabilitated per year). In that case, the model can be represented as follows (Figure 5.7).

Note that we have added a flow (bridge rehabilitated every year) and a converter (fraction of bridges rehabilitated [that require rehabilitation] every year). Now, the set of equations include two more parameters.

Fraction of bridges rehabilitated every year = 0.3 (adjustable)

Bridge rehabilitated every year = fraction of bridges rehabilitated every year (bridge/year).

Now, if we run the simulation, we get the results in Table 5.2. The effect of regular rehabilitation is obvious in the results.

## HOW DO WE CHARACTERIZE THE MODEL TRENDS?

More often than not we encounter trends in values of parameters (over time) that are nonlinear. Linear growth is very rare – that means there is no feedback from the system; what we view as linear is most likely an observation over a relatively short-time horizon. How can we physically explain them and model them? Common modes of dynamic behavior are presented below.

### EXPONENTIAL

Exponential behavior has been discussed above. Exponential behavior occurs when the net increase of the rate of the system is affected by the state of the system. Positive feedback leads to exponential growth, and the simplest system that can be envisioned is a first-order (containing one stock only) linear (equation of the flow is a linear

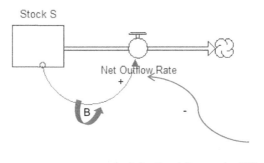

d, fractional decay rate, (1/time)

**FIGURE 5.8**   Feedback loop: balancing loop.

combination of variables) system. The net inflow is proportional to the state of the system, and this can be indicated as

$$\text{Net Inflow} = gS$$

where $g$ = fractional growth and $S$ = stock.

The behavior of the system can be indicated as

$$S(t) = S_0 e^{gt}$$

For the stock affecting the net **outflow**, the model will look like the one in Figure 5.8.

$$\text{Net Inflow Rate} = -\text{Net Outflow Rate} = -dS$$

$$\text{Average life of stock units} = 1/d$$

$$S(t) = S_0 e^{-dt}$$

Using an initial value of the stock $(S_0) = 1,000$ and a "d" value of 0.5, running the model generates data which can be plotted as shown in Figure 5.9, an example of exponential decay.

## GOAL SEEKING

In this case, the goal of the system (or the desired state of the system) can be thought of as a result of "corrective" action that determines the net inflow to the system, and the corrective action is a nonlinear function of the state of the system and the desired state of the system.

Mathematically, this can be expressed as

Net inflow = discrepancy between the state and the desired state of the system/AT, where AT is the adjusted time of time constant.

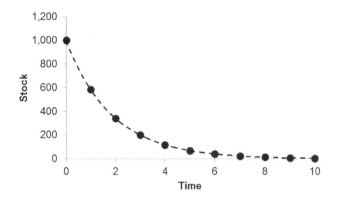

**FIGURE 5.9** Example of exponential decay.

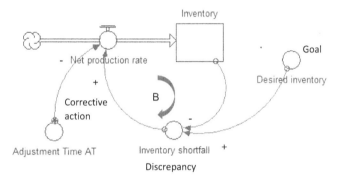

**FIGURE 5.10** Model of a goal-seeking system.

The model can be expressed as shown in Figure 5.10.

Using the above model, with an $S_0$ (initial inventory) = 100, desired inventory of 400, and AT = 2 (units of time), the plot of S versus time will be obtained as shown in Figure 5.11.

The equation can be written as follows:

$$S(t) = S* - (S* - S_0)e^{(-t/AT)}$$

where S* is the desired state of the stock (400 in the above example).

## OSCILLATION

In the case of a goal-seeking model, if there is any delay involved in say perceiving the shortfall in the inventory and taking the corrective action, then the resulting behavior of the stock will be of oscillating nature (over time). An example model is shown in Figure 5.12.

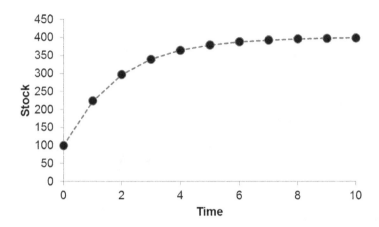

**FIGURE 5.11**    Results from simulations of the model of a goal-seeking system.

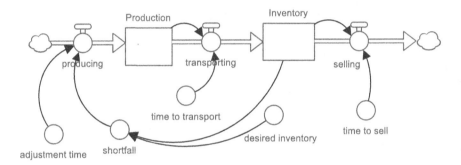

**FIGURE 5.12**    System dynamics model of an oscillating system.

Using the values of the parameter shown in Table 5.3, the plot of the inventory data from running the simulation versus time will look as shown in Figure 5.13.

## S-Shaped Growth

If there is a "limit" to the growth, then as a system approaches its limit, there is a nonlinear transition from a regime in which the positive feedback (self-reinforcing, indicated by R) dominates to a regime where the negative feedback (self-balancing, indicated by B) dominates. The behavior of such a system can be represented by an "S" curve or S-shaped growth or logistic growth. An example model is shown in Figure 5.14.

The parameters and equations are shown in Table 5.4. Note that the "damage" reinforces the rate of increase in damage, and the "fractional rate" works as the balancing parameter.

The plot of damage versus time is shown in Figure 5.15.

## TABLE 5.3
## Parameters and Values for Oscillating Stock

Top-Level Model:
Inventory(t) = Inventory(t − dt) + (transporting–selling) * dt
INIT Inventory = 5
INFLOWS:
transporting = Production/time_to_transport
OUTFLOWS:
selling = Inventory/time_to_sell
Production(t) = Production(t − dt) + (producing–transporting) * dt
INIT Production = 0
INFLOWS:
producing = shortfall/adjustment_time
OUTFLOWS:
transporting = Production/time_to_transport
adjustment_time = 2
desired_inventory = 10
shortfall = desired_inventory–Inventory
time_to_sell = 1
time_to_transport = 5
{ The model has 10 (10) variables (array expansion in parens).
In root model and 0 additional modules with 0 sectors.
Stocks: 2 (2) Flows: 3 (3) Converters: 5 (5)
Constants: 4 (4) Equations: 4 (4) Graphicals: 0 (0)
}

**FIGURE 5.13** Plot of stock versus time from model of an oscillating system.

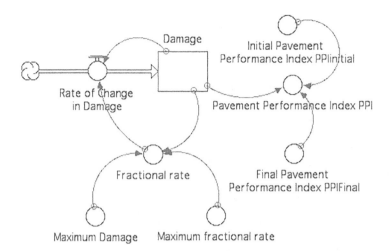

**FIGURE 5.14**    Model of an S-shaped growth.

## TABLE 5.4
### Parameters and Equations for Model of S-Shaped Growth

Damage(t) = Damage(t – dt) + (Rate_of_Change__in_Damage) * dt
INIT Damage = 1e-06
INFLOWS:
Rate_of_Change__in_Damage = Damage*Fractional_rate
Final_Pavement_Performance_Index_PPIFinal = 20
Fractional_rate = Maximum_fractional_rate*(1 – Damage/Maximum_Damage)
Initial_Pavement_Performance_Index_PPIinitial = 100
Maximum_Damage = 1
Maximum_fractional_rate = 1
Pavement_Performance_Index_PPI = Initial_Pavement_Performance_Index_PPIinitial-Damage*(Initial_
  Pavement_Performance_Index_PPIinitial-Final_Pavement_Performance_Index_PPIFinal)

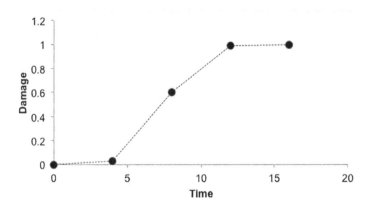

**FIGURE 5.15**    Plot of damage from S-shaped growth model.

## MODELING WITH A SPREADSHEET

Note that it is possible to construct relatively simple models in a spreadsheet, such as in MS Excel. Let us take an example. Suppose a city decides to maintain a specific inventory (number of kms, 100) of roads in maintained (good) condition at any time. At the time of this decision, the city has 20 kms of road in maintained condition. Hence, there is a shortfall, that is, the difference between desired inventory and available inventory. It takes 2 months to complete the bidding process for any maintenance job and another 2 months to mobilize equipment and personnel to carry out maintenance procedures for each kilometer of the road network. At the same time, it takes 24 months for a stretch of a roadway to deteriorate to a condition where it again needs maintenance. We want to model this problem and simulate it to find out how the inventory of maintained roads changes over time. Table 5.5 shows the cells

**TABLE 5.5**
**Excel Spreadsheet Model**

| | A | B | C | D | E | F | G | H |
|---|---|---|---|---|---|---|---|---|
| 1 | Desired inventory of maintained roads, kms | 100 | | | | | | |
| 2 | Time to bid, months | 2 | | | | | | |
| 3 | Time to mobilize and construct, months per km | 2 | | | | | | |
| 4 | Average time to deteriorate per km, months | 24 | | | | | | |
| 5 | | | | | | | | |
| 6 | Time in months | **0** | **1** | **2** | **3** | **4** | **5** | **6** |
| 7 | Beginning bid roads, kms | 0 | 0 | 40 | 50 | 34 | 17 | 8 |
| 8 | Shortfall, kms | 80 | 80 | 61 | 17 | −21 | −42 | −49 |
| 9 | Bids, per month | 0 | 40 | 30 | 9 | 0 | 0 | 0 |
| 10 | Mobilizing, constructing, per month | 0 | 0 | 20 | 25 | 17 | 8 | 4 |
| 11 | Ending bid, kms | 0 | 40 | 50 | 34 | 17 | 8 | 4 |
| 12 | | | | | | | | |
| 13 | Beginning inventory of maintained roads, items | 20 | 20 | 39 | 83 | 121 | 142 | 149 |
| 14 | Mobilizing, constructing, per month, per month | | 0 | 20 | 25 | 17 | 8 | 4 |
| 15 | Deteriorating, per month | | 1 | 2 | 3 | 5 | 6 | 6 |
| 16 | Ending inventory of maintained roads, kms | 20 | 19 | 58 | 105 | 133 | 144 | 147 |

*Note:* Data up to 6 months are shown; the rest of the columns have the same formula for the cells; copy Column C data and paste for the other months (Month 2 onward).

**Explanation of Cells**
Rows 1–4: Constants, as explained in Column A
Row 6: Time in months; B6 = 0, C6 = (B6 + 1)
Row 7: Beginning bid roads, kms; B7: starting at 0 year, 0; C7 = B11
Row 8: Shortfall (between desired and ending inventory): B8: ($B$1 − B13); C8: ($B$1 − C13)
Row 9: Bids, per month − B9: 0 for 0th year, C9: (Shortfall/Time to bid): C8/$B$2
Row 10: B10 = B7/$B$3, C10 = C7/$B$3
Row 11: B11 = B7 + B9 − B10, C11 = C7 + C9 − C10
Row 13: B13 = 20; C13 = B16 + C10
Row 14: B14: 0; C14 = C10
Row 15: B15 = 0; C15 = C13/$B$4
Row 16: B16 = 20; C16 = C13 + C14 − C15

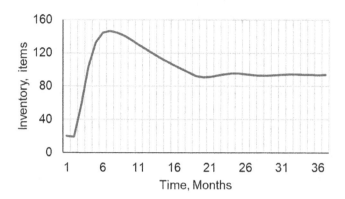

**FIGURE 5.16**    Plot of inventory of maintained roads versus time from the spreadsheet model.

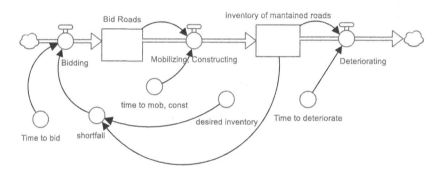

**FIGURE 5.17**    System dynamics model in STELLA.

with explanations, and Figure 5.16 shows a plot of inventory (maintained roads, kms) over time. The constants, shown in the first four rows of Table 5.5, are equivalent to converters; the parameters in rows 7, 10, 14, and 15 are equivalent to flows; and the rest are equivalent to stocks. Note that the results show an oscillatory behavior at first and then tend to stabilize.

For practical purposes, it is advisable to use dedicated software for modeling system dynamics problems and not use spreadsheets. One reason is that it becomes difficult, if not impossible, to indicate and represent the feedback loops, and another reason is that the model cannot be visually represented in the spreadsheet as explicitly as in a dedicated software such as Stella or Vensim.

The system dynamics model in STELLA is shown in Figure 5.17.

# 6 Cascading Effects

Let us present two principles that will be explained through a model in this chapter. The important points about these two principles are feedback loops and their disproportionate effect.

*Principle 1.* **Feedback** *loops cause* **cascading** *damage to pavement layers and adjacent sections and the pavement system; Principle 2. Damages caused to the pavement system by feedback loop(s) are* **not proportional** *to damages in individual layers or adjacent sections*

For an explanation of these principles, consider this example. A stretch of roadway pavement consists of many adjacent sections, which are constructed at different levels, at different times, with different batches of materials during the same construction phase, and may end up with different quality of construction. However, when we consider the performance of the stretch of the roadway, or a specific section (such as a test section), we do not consider the impact of the adjacent sections. For example, we also do not consider that water entering through a poorly constructed joint or a section in an adjacent lane can enter another section at a lower level (in most cases, pavements are sloped longitudinally and laterally for drainage) and cause damage (see Figure 6.1).

An adjacent section with low density/cracks and high permeability can lead to such a consequence; furthermore, two adjacent sections can be in two different qualities, and the performance of the inferior section will negatively affect performance of the superior section. As a result, both sections will deteriorate at a rate that is higher than expected for a single section. The point is that the deteriorations of sections are dependent on each other and that the performance of each section cannot be predicted with accuracy without considering the contributions of damage from other sections. Particularly, if water can move freely between adjacent sections, there is a cascading effect of the deterioration of sections into other adjacent sections. For a stretch of a roadway, if the entire set of sections is not considered as a system and analyzed, there will be an overestimation of the performance of a specific section. How can we analyze this problem with system dynamics?

Let us take the following steps in the modeling and analyses (Mallick and Kautz, 2019).

Consider two adjacent pavement sections, making up a "job", that are isolated from other parts of the pavement. Consider the sections to have an initial damage of (1 − quality), where quality can range from 0 to 1, on the basis of as constructed in-place density. Consider a logistic model for the variation of damage over time.

$$D = \frac{1}{1 + e^{-t}}$$

The concept of the model is shown in Figure 6.2. Note the feedback included in the model.

DOI: 10.1201/9781003345596-6

**FIGURE 6.1**    Water entering through a poorly constructed joint has a damaged adjacent lane (patched) (Courtesy, Mr. P. S. Kandhal).

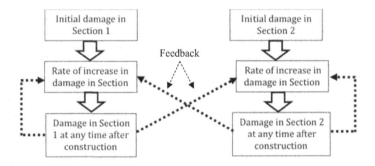

**FIGURE 6.2**    Concept of the model.

The equations are shown in Table 6.1. There is a feedback loop from damage (a stock in system dynamics nomenclature) to the rate of damage (a flow), and the rate of damage is indicated as a function of the current damage level and a fractional rate. The effect of the current damage, the maximum damage, and a maximum fractional rate is considered in the fractional rate to determine the rate of change in damage of a section.

Without any effect of Section 2 on Section 1, Rate of Change in Damage in Section 1

$$= (\text{Damage in Section 1}) * (\text{Fractional rate in Section 1})$$

With the effect of Section 2 on 1, Rate of Change in Damage in Section 1

$$= (\text{Damage in Section 1}) * (\text{Fractional rate in Section 1})$$

$$* (\text{Factor of Section 1 dependent on damage of Section 2})$$

## TABLE 6.1
## Equations in the System Dynamics Model

Damage_in_section_1(t) = Damage_in_section_1(t − dt) + (Rate_of_Change__in_Damage_in_
section_1) * dt

INITIAL Damage_in_section_1 = 0.0001

INFLOWS:

Rate_of_Change__in_Damage_in_section_1 = Damage_in_section_1*Fractional_rate_1*Factor_of_
section_1_dependent_on_damage_of_section_2

Damage_in_section_2(t) = Damage_in_section_2(t − dt) +
(Rate_of_Change__in_Damage_in_section_2) * dt

INIT Damage_in_section_2 = 0.0001

INFLOWS:

Rate_of_Change__in_Damage_in_section_2 = Damage_in_section_2*Fractional_rate_2*Factor_of_
section_2_dependent_on_damage_of_section_1

Factor_of_section_1_dependent_on_damage_of_section_2 = 3.9*(Damage_in_
section_2^2)+0.17*Damage_in_section_2+0.9

Factor_of_section_2_dependent_on_damage_of_section_1 = 3.9*(Damage_in_
section_1^2)+0.17*Damage_in_section_1+0.9

Fractional_rate_1 = Maximum_fractional_rate_in_section_1*(1 − (Damage_in_section_1)/
Maximum_Damage_in_section_1)

Fractional_rate_2 = Maximum_fractional_rate_in_section_2*(1 − (Damage_in_section_2)/
Maximum_Damage_in_section_2)

Maximum_Damage_in_section_1 = 1

Maximum_Damage_in_section_2 = 1

Maximum_fractional_rate_in_section_1 = 1

Maximum_fractional_rate_in_section_2 = 1

where *Factor of Section 1 dependent on damage of Section 2 = 3.9\*(Damage in section 2^2) + 0.17\*Damage in Section 2 + 0.9,*
where the equation is from the following hypothetical data.

| Damage of Section 2 | Factor of Section 1 Dependent on the Damage of Section 2 |
|---|---|
| 0.1 | 1 |
| 0.5 | 2 |
| 1 | 5 |

Now simulate the model for a time of 20 years, and plot the results of damage versus time.

See the four plots in Figure 6.3 for the different cases of connections and initial quality. The connection and the existence of different qualities result in quicker deterioration – note that not considering this effect leads to overestimation of the life of the pavement sections.

Figure 6.3 shows that if we consider the sections to be disconnected and at the same initial condition, the damage versus time plots for the two sections overlap each

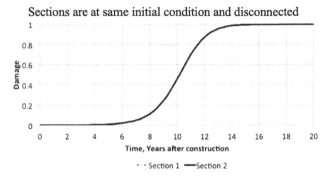

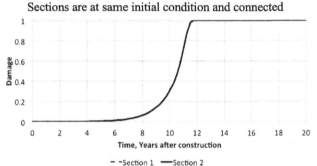

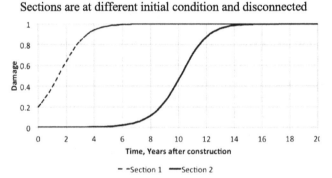

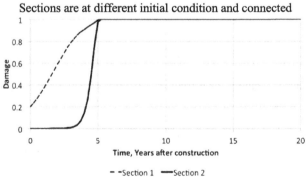

FIGURE 6.3    Damage versus time plots.

other, it takes 14 years for both sections to reach a fully damaged state (Damage = 1). If the sections are connected and at the same initial condition, the plot overlaps each other but the life (time to reach a fully damaged state) is reduced, it is now less than 12 years. This shows the cascading effect of damage in adjacent sections. Now, if the two sections are at different initial conditions and not connected, their plots do not overlap, and Section 1 reaches a fully damaged state much earlier (6 years) than Section 2 (14 years). However, if they are connected, their rates of damage over time are different (the initial states are different), and both of them tend to reach a fully damaged state in 5 years – less than either of the two durations (6 and 14 years for the two sections) noted above. This shows that if we consider a pavement system, with Sections 1 and 2, and consider the failure (fully damaged state) of the system as failure in either or both of the sections, we cannot predict analytically what is its duration, based on the failure duration of Sections 1 and 2 – there is no proportional effect. The effect is evident only from the simulation of the model. How can the results help us? We can use the results to identify the most important/vulnerable section and quality control specifications.

Extend the model to a six-section pavement system (Figure 6.4), where the slope of the pavement is from top to bottom and left to right, and assume that the section at a lower elevation is affected by an adjacent section at a higher elevation. The slopes are considered as indicators of water flow inside the sections. In this case, it

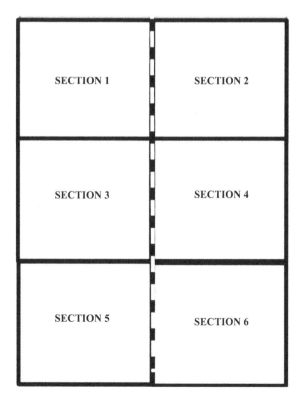

**FIGURE 6.4**  Layout of six sections.

is assumed that infiltrated water will flow from left to right, and as expected, from higher to lower elevation. This is a practical example, which is likely to be encountered in most, if not all, pavement construction projects.

Now simulate the model first by assuming that all the sections are at the same initial state of damage (same quality at the end of construction). The results of simulation (Figure 6.5) show that the different sections reach damaged state at different times, which range from 3 to 8 years. Section 6, which is at the lowest elevation and on the right side, reaches damaged state in 3 years – much earlier than Section 1, which takes 8 years. The damage versus time plots for other sections lie in-between those of Sections 1 and 6. Note that while Section 6 is affected by water flow from all other sections, Section 1 is not affected by any. Since Section 2 is affected by only one section (1), its damage plot is not significantly different from that of Section 1 (they overlap); however, other sections, which are affected by more than two sections, reach the damaged state much faster, all in less than 5 years. Hence, if the six sections constitute a pavement system or pavement construction project, one can expect it to fail (full damage), in at least one section in 3 years, and all but two sections in less than 5 years. This fast rate of failure, or shortened life, can be a "surprise" if all the sections were designed for a specific life, say 10 years.

Now simulate the model with the mean damage of the first section varying from 0.005 to 0.2. The results (Figure 6.6) show the impact of the initial quality of Section 1 on the life of other sections. Being at the highest elevation and left-most corner, the impact of this section's initial quality is significant – for a range of initial damage of 0.005–0.2, other sections reach damaged state within 3–6 to 2.5–3 years. This simulation points out the importance of the identification of the critical section in a system (in this case, it is Section 1). This helps us to take extra precautions and apply stricter quality control while constructing the critical section. If this section is constructed with high density, at a very low initial damaged state, the life of at least few other sections can be significantly prolonged. Similar argument applies for Section 2. Similarly, we can also see which sections are the most vulnerable, and it is important to take extra precautions for these sections as well. For example, these sections can

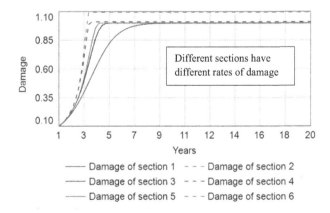

**FIGURE 6.5** Plots of damage versus time with sections with the same initial quality.

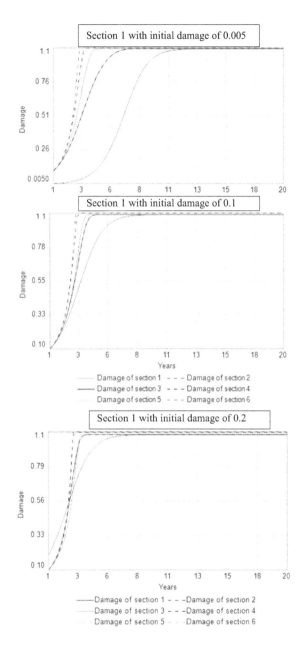

**FIGURE 6.6** Plots of damage versus time with the mean damage of the first section varying from 0.005 to 0.2.

be constructed with highly moisture-resistant materials, with the expectation that water will flow from the sections at higher elevations, and that the highly moisture-resistant materials will protect the section from the water. The bottom line is that wherever ingress of water through the surface and flow of water inside the pavement are expected, improvement of pavement quality will enhance the life of the sections, and hence that of the entire pavement system.

Next, simulate the model with the damage of all sections expressed as normally distributed with a mean of 0.05 and standard deviation of 0.1; see the results of simulations in Figure 6.7.

This case is more realistic, since some variation can always be expected in construction projects. What is very clear is that the range of life increases as we move downstream. It is between 8–9, 5–9, 5–9, 4–9, 4.5–8, and 4–9 for Sections 1, 2, 3, 4, 5, and 6, respectively. This points out the fact that variability in the quality of Section 1 can have a significant effect on the variability of life in other sections, and high variability means a corresponding low reliability for the life of other sections. Therefore, in addition to achieving a target of high density, say, for Section 1, it is important to reduce its variability also. In fact, good quality does imply being both accurate (on target) and precise (low variability).

Now what if we say that the first section impacts the second only after the first section has undergone a certain damage (threshold, e.g., significantly cracked to let water in), sections at lower levels will be affected more, and the impact is dependent on the quality of the material of the section. Simulate the model, with the following conditions (see the additional lines in Table 6.2). When the damage in Section 1 reaches 0.6, the impact factor (impacting the damage to another section) starts acting (is triggered). There is a delay for impact factor of sections at each elevation,

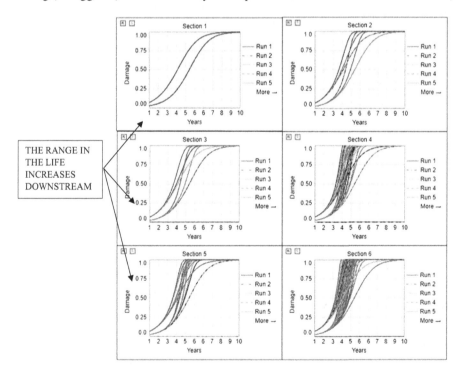

**FIGURE 6.7**   Plots of damage versus time model with the damage of all sections expressed as normally distributed with a mean of 0.05 and standard deviation of 0.1.

## TABLE 6.2
## Additional Equations for Incorporating Effects of Delay and Quality of Recipient Section

Impact_of_sections[1] = IF (Damage_of_section[1]>0.6) THEN ((((3.9*Damage_of_ section[1])^2) + (0.17*Damage_of_section[1]) +1) ELSE 1
Impact_of_sections[2] = IF (Damage_of_section[2]>0.6) THEN ((((3.9*Damage_of_ section[2])^2) + (0.17*Damage_of_section[2]) +1) ELSE 1
Impact_of_sections[3] = IF (Damage_of_section[3]>0.6) THEN ((((3.9*Damage_of_ section[3])^2) + (0.17*Damage_of_section[3]) +1) ELSE 1
Impact_of_sections[4] = IF (Damage_of_section[4]>0.6) THEN ((((3.9*Damage_of_ section[4])^2) + (0.17*Damage_of_section[4]) +1) ELSE 1
Impact_of_sections[5] = IF (Damage_of_section[5]>0.6) THEN ((((3.9*Damage_of_ section[5])^2) + (0.17*Damage_of_section[5]) +1) ELSE 1
Impact_of_sections[6] = IF (Damage_of_section[6]>0.6) THEN ((((3.9*Damage_of_ section[6])^2) + (0.17*Damage_of_section[6]) +1) ELSE 1
DELAY1(Impact_of_sections[1], 10*(Material_Quality[2])/(Mean_Elevation[1]-Mean_Elevation[2]))
Time_rate_of_increase_of_damage_in_section[3] = Fractional_rate_in_section[3]*Damage_of_ section[3]* DELAY1(Impact_of_sections[1], 10*(Material_Quality[3])/ (Mean_Elevation[1]-Mean_Elevation[3]))
Time_rate_of_increase_of_damage_in_section[4] = Fractional_rate_in_section[4]*Damage_of_ section[4]* (DELAY1(Impact_of_sections[3], 10*(Material_Quality[4])/(Mean_Elevation[3]-Mean_ Elevation[4])) + DELAY1(Impact_of_sections[2], 10*(Material_Quality[4])/ (Mean_Elevation[2]-Mean_Elevation[4])) )
Time_rate_of_increase_of_damage_in_section[5] = Fractional_rate_in_section[5]*Damage_of_ section[5]* DELAY1(Impact_of_sections[3], 10*(Material_Quality[5])/ (Mean_Elevation[3]-Mean_Elevation[5]))
Time_rate_of_increase_of_damage_in_section[6] = Fractional_rate_in_section[6]*Damage_of_ section[6]* (DELAY1(Impact_of_sections[5], 10*(Material_Quality[6])/(Mean_Elevation[5]-Mean_ Elevation[6])) + DELAY1(Impact_of_sections[4], 10*(Material_Quality[6])/ (Mean_Elevation[4]-Mean_Elevation[6])))

depending on its elevation. The impact on the recipient section is dependent on the quality of the recipient section.

The model was run with random initial damages (quality = 1 – damage), material quality, and mean elevations as shown in Table 6.3.

## TABLE 6.3
## Initial Condition, Material Quality, and Mean Elevation of the Six Sections

| Section | Initial Quality | Material Quality | Mean Elevation |
|---|---|---|---|
| 1 | 0.99 | 0.9 | 1 |
| 2 | 0.89 | 0.6 | 0.95 |
| 3 | 0.92 | 0.7 | 0.85 |
| 4 | 0.85 | 0.9 | 0.8 |
| 5 | 0.93 | 0.8 | 0.78 |
| 6 | 0.88 | 0.9 | 0.75 |

The results are shown in Figure 6.8. Note that while Section 1 lasts for 12 years, Sections 4 and 6, which are on the right and bottom-most (water flows from left to right and top to bottom), will last for only 4 years. This clearly identifies those sections as the most vulnerable ones and may require stricter quality control or better materials (may not be practical) during construction. What happens if all the sections are at the same initial quality?

Now, run the model with all the sections at the same initial quality (0·99) and the same material quality (0·9) but with different elevations (Table 6.2); see the results in Figure 6.9. Note that due to an increase in the initial quality, Sections 4 and 6 last a bit longer (5 years instead of 4 from the last case), but they still remain the most vulnerable sections. Will it help if these sections have higher initial qualities?

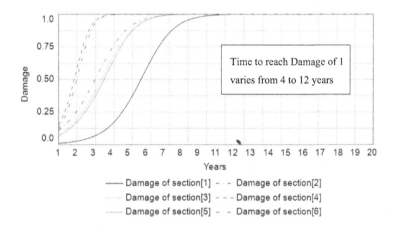

**FIGURE 6.8** Plots of damage versus time with delay with a randomly selected set of initial qualities (damage = 1 − quality) and qualities (Tables 6.2 and 6.3).

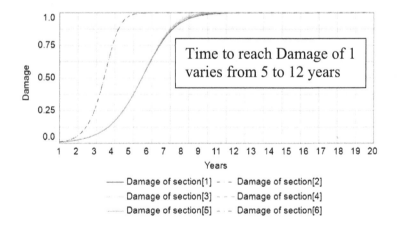

**FIGURE 6.9** Plots of damage versus time with all the sections at the same initial quality (0.99) and the same material quality (0.9) but with different elevations (Tables 6.2 and 6.3).

Now, run the model by raising the initial quality of Sections 4 and 6 (0.9998, instead of 0.99); see the results in Figure 6.10. The increase in lives due to the increased initial qualities for Sections 4 and 6 is clearly visible – both sections now last for 7 years, much closer to the 10 years life span of Section 1. However, these sections still remain the most vulnerable sections.

Next, let us calculate the expected elevations and use them in the model. The elevation of the outer and inner lane can be set at 0.75 and 1 m, respectively, considering a 2% cross slope, and those for the different sections can be considered as shown in Table 6.4.

Now simulate the model with values of initial conditions, material quality (Table 6.3), and mean elevation (Table 6.4). See the results in Figure 6.11. Again, it can be seen that Sections 4 and 6 reach the end of life within 5 years, whereas other sections take 9–12 years. Considering the six sections as a system or a project raises an important question: what is the life of this project? When will it need rehabilitation? Since it is impractical to consider rehabilitation/construction work for two sections out of six sections, do we wait until, say, 10 years? If that is the case, then for the last

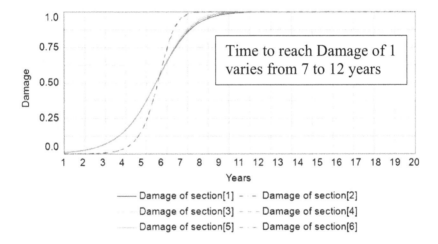

FIGURE 6.10    Plots of damage versus time with raising the initial quality of Sections 4 and 6.

TABLE 6.4
Calculated Elevations of
the Different Sections

| Section | Mean Elevation |
|---------|----------------|
| 1 | 1.00 |
| 2 | 0.75 |
| 3 | 0.75 |
| 4 | 0.50 |
| 5 | 0.50 |
| 6 | 0.25 |

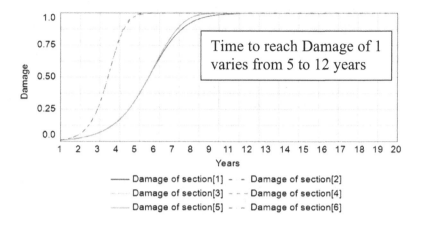

FIGURE 6.11   Plots of damage versus time for model with actual values of initial conditions, material quality, and mean elevation, calculated as follows. The elevation of the outer and inner lanes can be set at 0.75 and 1 m, respectively, considering a 2% cross slope.

FIGURE 6.12   Water and distresses on one lane of the road, while the other side remains relatively dry and intact.

5 years, motorists will have a very challenging time traveling through Sections 4 and 6, which are on the same side of the road. This scenario is actually very common in real life, where one side is relatively intact, and the other side clearly shows water infiltration and related distresses (see Figure 6.12).

So what happens if we try to rectify this situation by improving the initial quality of Sections 4 and 6? To explore this scenario, simulate the model with same materials and initial quality of 0.9900, except 4 and 6, for which the initial quality is 0.9999 considering a 2% grade in both long and lateral directions. See the results in Figure 6.13. Instead of a difference of 5–7 years between the lives of the different sections, now we have a relatively smaller gap of 2 years (8–10 years) between the fully damaged states of the different sections. This translates to a significant reduction in time for which travel could be unsafe and uncomfortable. This also means a significant increase in the time interval after which the pavement agency has to consider construction work – savings in cost, materials, and energy.

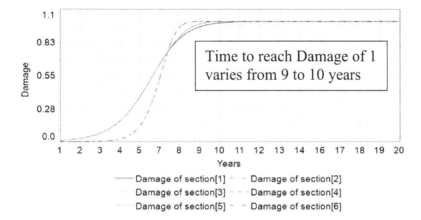

**FIGURE 6.13** Plots of damage versus time for model with the same materials and an initial quality of 0.9900, except 4 and 6, for which the initial quality is 0.9999 considering 2% grade in both long and lateral directions.

Now, since we are able to model the problem and show that an improvement in the initial quality of certain sections can homogenize the life of the different sections, let us experiment with different scenarios and find out what are the required parameters to ensure similar life or rate of deterioration for all the section. The conditions and the results are in Table 6.5. Instead of having stricter quality control and

**TABLE 6.5**
**Case Study Simulation Results**

| Case | Difference in Elevations | The "Trigger" Damage of the "Contributing" Sections | Initial Quality of the Contributing Sections (1, 2, 3, 5) | Required Initial Quality of Recipient Sections (4, 6) (To Make Them Cross 0.6 at the Same Year and Reach 1 within 1 Year of the Others) | Difference between Desired Initial Quality of Recipient Sections and Initial Quality of Contributing Sections (%) |
|---|---|---|---|---|---|
| 1 | 0.25 | 0.6 | 0.9 | 0.99 | 10 |
| 2 | 0.10 | 0.6 | 0.9 | 0.99 | 10 |
| 3 | 0.05 | 0.6 | 0.9 | 0.99 | 10 |
| 4 | 0.25 | 0.5 | 0.9 | 0.99 | 10 |
| 5 | 0.25 | 0.4 | 0.9 | 0.99 | 10 |
| 6 | 0.25 | 0.3 | 0.9 | 0.99 | 10 |
| 7 | 0.25 | 0.6 | 0.8 | 0.95 | 18.8 |
| 8 | 0.25 | 0.6 | 0.85 | 0.97 | 14.1 |
| 9 | 0.25 | 0.6 | 0.95 | 0.99 | 4.9 |

higher quality of the recipient sections (4 and 6), it will be more practical to use uniform quality control, but with a higher quality overall. The last column in Table 6.5 shows that as the initial quality of the contributing sections increases, the difference between the desired initial quality of recipient sections and the initial quality of contributing sections (%) decreases. Therefore, the utility of this type of modeling and simulation is that in cases of cascading and disproportionate effects, we can estimate the controllable parameters (lever points) that can help us achieve desirable results.

## REFERENCE

Mallick, R. B. and Kautz, F. A. (2019). Effect of quality on damage in contiguous pavement sections. *Journal of Infrastructure Asset Management*, 6(4), 233–244. doi: 10.1680/jinam.18.00047

# 7 Feedback Causal System

In this chapter, the following principle is presented with an example, where the main emphasis is on the demonstration of *cascading* effects. Principle 3. **Cascading** effects in different parts of a pavement system are caused by changes in the parts of the system and their relationships.

The pavement that is of interest in this chapter is a common type – a relatively thin asphalt mix layer over a granular base, typically found in low- to medium-traffic-volume roads all over the world. Moisture-damaged surface layers and cracked pavements with potholes are common occurrences in these types of pavements (Figure 7.1).

When a surface crack develops in the asphalt mix, water enters into the granular layer, raises its saturation level, reduces its stiffness, and deteriorates its structural capacity. As a result, the structural capacity of the entire pavement decreases, leading to further surface cracks (due to excessive deformation under traffic), which leads to further water ingress. This results in a vicious cycle (reinforcing loop with cascading effect, Figure 7.2) of cracking and water infiltration, and a cascading effect of one layer upon the other, which is typically not considered in pavement analysis, although research based on long-term pavement performance test sections has conclusively shown (Haider et al., 2019) an increase in base moisture content with an increase in the amount of surface cracking. In system dynamics, this type of problem, which is not apparent, unless modeled explicitly and analyzed with a feedback loop, is identified as one having a "latent" (Saeed, 2015) structure.

Let us construct a system dynamics model to explore this problem and determine whether paving fabrics to restrict flow from the surface to the base or sealing of cracks have any beneficial effects (Mallick, 2021).

Develop the concept/causal map of the model (Figure 7.3). In this step, based on the problem illustrated in Figure 7.2, write down the factors and parameters in a logical fashion and sequentially, with appropriate connections, which are responsible for the problem. The feedback loop should be explicitly shown here. In this problem, the starting point is the two critical surface layer properties, thickness and permeability, as they govern the ingress of water into the pavement structure. Additionally, permeability through cracks, if any, should also be considered. Next, the thickness of the granular base and its permeability are considered, as they dictate the time it takes for the base to get fully saturated. As the base saturation changes, the resilient modulus of the base changes – hence, it should be considered next. Following this, the tensile strain at the bottom of the surface asphalt mix layer and the resultant fatigue life should be included in the model. To calculate the damage of the surface asphalt mix layer at any time, along with the fatigue life, the actual traffic loading should be considered. The damage of the base can be calculated with respect to the ratio of the actual resilient modulus to the design modulus. Finally, the feedback loop between the damage of the base and the damage of the surface asphalt layer must be indicated.

DOI: 10.1201/9781003345596-7                                   **49**

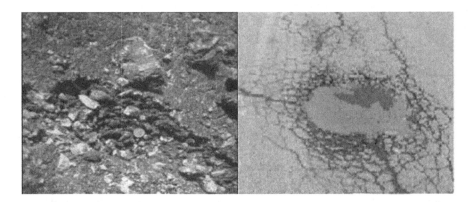

**FIGURE 7.1**    Moisture damage in asphalt mix and fatigue cracking.

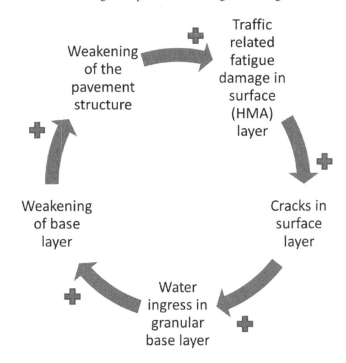

**FIGURE 7.2**    Reinforcing loop of water infiltration and cracking.

Next, prepare a system dynamics model for the cumulative change in damage due to traffic (fatigue) and moisture damage (see Figure 7.4 for the model and Table 7.1 for the equations). The base and surface damage constitute the two main stocks, and the two main flows are their corresponding rates. Note that this model has two "sectors" – surface damage and base damage. The grouping of parts of the model, which pertain to one problem, in separate sectors, helps in getting visual clarity and allows the user to run parts of the model (one sector) without having to run the other parts. This helps in identifying problems, if any, in parts of the model, before it becomes too

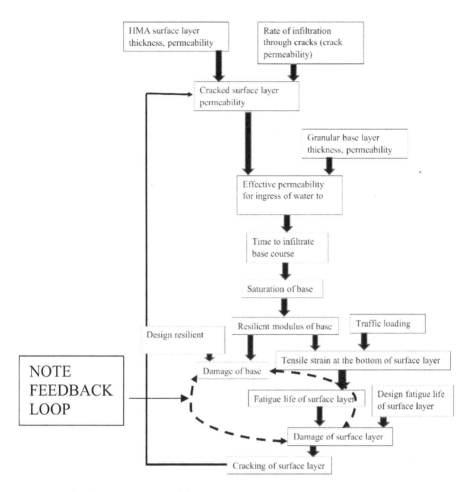

**FIGURE 7.3** Concept of the model.

complicated to catch any error in the entire model. Once each sector is found to be working, the different sectors can be linked, and the entire model can be simulated.

Now provide estimates and/or equations for the parameters of the model (Table 7.2). This step can be a significant one. Remember that some of these parameters can be from the pertinent literature, while others can be reasonably assumed on the basis of experience and/or experiments. In this problem, you will notice that the relevant source is indicated in the last column of Table 7.2. In some cases, you may need to consult relevant documents for figures or equations, from which you need to derive or calculate the estimates for the required parameters. Examples of such steps are shown for this problem.

Select a stiffness versus saturation model for the base course; for stiffness, select resilient modulus; see Figure 7.5.

$$M_R = M_{Ropt}\left(-6.4171S_R^4 + 14.92S_R^3 - 9.8308S_R^2 - 0.011S_R + 1.8264\right)$$

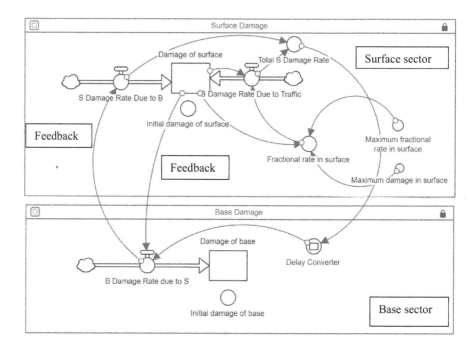

**FIGURE 7.4**   System dynamics model with two sectors – surface and base damage.

where $M_{Ropt}$ = optimum or design resilient modulus of the granular base, $M_R$, MPa, $S_R$ = saturation.

Next, calculate the surface damage corresponding to base damage, with 20% of surface cracking considered to be 100% failure or damage. A 30-minute (0.5 hours) rainstorm is considered for this problem. Derive an equation of base damage as a function of surface damage. The procedure is explained below (Figure 7.6). The derivative of this equation is then used to calculate the rates of damage.

Now, derive an equation of surface damage as a function of the base damage. The procedure is explained below (Figure 7.7). The derivative of this equation is then used to calculate the rates of damage.

In this step, the resilient modulus of the base was used along with 3,500 MPa dynamic modulus of HMA and 50 MPa resilient modulus of the subgrade (load of 40 kN with a contact area of 0.058 m²) to calculate the tensile stain at the bottom of the HMA layer in layered elastic analysis. This strain was then used to calculate the repetitions to failure using the Asphalt Institute fatigue cracking formula.

$$N_f = (18.4)(0.00432)(C)\left(\varepsilon_t^{-3.291}\right)\left(\frac{1,000\,\text{Modulus}}{6.9}\right)^{-0.854}$$

where $N_f$ = repetitions to failure; 18.4 = shift factor; $C = 10^M$; $M = 4.84*((V_b/(V_a + V_b)) - 0.69)$, $\varepsilon_t$ = tensile strain at the bottom of the surface HMA layer; Modulus = dynamic modulus of HMA, MPa; $V_a$ = Air Voids, 5%; $V_b$ = Binder content by volume, 13.63%.

# TABLE 7.1
## Model Equations

| Sector | Parameter | Equation |
|---|---|---|
| Base damage | Damage of base (t) | Damage of base(t − dt) + (Base Damage Rate due to Surface) * dt |
| | Initial damage of base | Specified initial damage of base (0.05) |
| | INFLOW: B damage rate due to S | Delay converter*((−3.85*4*Damage_of_surface^3) + (11.04*3*Damage_of_surface^2) − (11.69*2*Damage of surface) + 5.45) (Deal converter delays the change in value according to the input, such as SMTH1, see the next row) |
| | Delay converter | SMTH1(Total S damage rate, 5, 0) (SMTH1 calculates a first-order exponential smoothing of input, using an exponential averaging time of average time (5 years, in this case), and an initial value for the smoothing (0, in this case)). |
| Surface damage | Damage of surface (t) | Damage of surface (t − dt) + (S damage rate due to traffic + surface damage rate due to base) * dt |
| | Initial damage of surface | Specified initial damage of surface (0.05) |
| | Inflow (rate): surface damage rate due to traffic | Fractional rate in surface*damage of surface |
| | Inflow (rate): surface damage rate due to base | 1.114*Base damage rate due to surface |
| | Fractional rate in surface | (Maximum_fractional_rate_in_surface)*(1 − (Damage_of_surface)/Maximum_damage_in_surface) |
| | Maximum damage in surface | 1 |
| | Maximum fractional rate in surface | Specified (0.4) |
| | Total surface damage rate | Surface damage rate due to base + surface damage rate due to traffic |

*Note:* (t) indicates a function of time; d(t) = 1/16 of time interval; time interval = 1 Year; Euler integration method applied for calculation of the damages from the inflows (rates).

## TABLE 7.2
## Parameters Used in the Model

| Parameter | Value/Equation | Source |
|---|---|---|
| Uncracked HMA permeability, $k_{uncracked}$, m/s | $10^{-7}$ | For 9.5 mm fine-graded mix, Brown et al. (2008) |
| Rate of infiltration through cracks, $k_{cracked}$, m/s | $2.06 \times 10^{-4}$ | Ridgeway (1976) |
| Permeability of cracked HMA, $k_{HMA}$, m/s | $(1 - \%\text{Area cracked}/100) * k_{uncracked}$ HMA $+ (\%$ area cracked$/100) * k_{cracked}$ HMA | NA |
| Thickness of base, $h_{Base}$, m | 0.6 | NA |
| Thickness of HMA (surface), $h_{HMA}$, m | 0.15 | NA |
| Diameter corresponding to 10% passing for base, D10, mm | 0.3 | Assumed |
| Maximum density, $t/m^3$ | 2.5 | Assumed |
| Density, $t/m^3$ | 1.6 | Assumed |
| Porosity, n | $1 - \dfrac{\text{Density}}{\text{Maximum Density}}$ | NA |
| Percent finer than 0.075 mm, $P_{0.075}$ | 5 | Assumed |
| Permeability of base, $k_{Base}$, m/s | $2.192*(\text{D10, mm}^{1.478})*(\text{porosity}^{6.654})/$ (percent finer than $0.075\text{mm}^{0.597}$) | Main Roads (2003) |
| Effective permeability through surface and base, $k_{effective}$, m/s | $k_{effective} = \dfrac{h_{HMA} + h_{Base}}{\dfrac{h_{HMA}}{k_{HMA}} + \dfrac{h_{Base}}{k_{Base}}}$ | NA |
| $\theta_s$ = Volumetric moisture content at saturation | n-0.05 | NA |
| Optimum moisture content (OMC) of base, % | 7 | Assumed |
| Annual rainfall, AR, m | 1.1 | Assumed |
| Potential evaporation, PE, m | 0.2 | Assumed |
| Percent finer than 0.425 mm, $P_{0.425}$, mm | 15 | Assumed |
| $\theta_i$ = Initial volumetric moisture content | $0.70*(\text{OMC})+0.29*(\text{AR-PE})+0.58*(P_{0.425}/ P_{0.075}) - 0.02*P_{2.36}$ | |
| $L_f$ = Total thickness of infiltration, m | $(h_{HMA} + h_{Base})$ | NA |
| $\Psi_f$ = Suction, m | 0.1 | Assumed |
| $h_L$ = Depth of ponded water, m | 0.0254 | Assumed |
| Time to infiltrate, t, s | $\dfrac{\theta_s - \theta_i}{k_{effective}}\left[ L_f - \left(h_L - \psi_f\right)\ln\left(\dfrac{h_L + L_f - \psi_f}{h_L - \psi_f}\right)\right]$ | Green and Ampt (1911) |

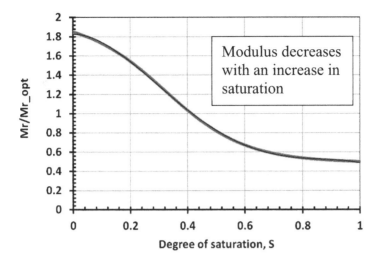

**FIGURE 7.5**    Plot of degree of saturation versus resilient modulus of granular base.

Next, model the damage growth due to traffic-related fatigue cracking as a logistic growth that is typically used for characterizing damage accumulation in pavements (ARA, 2004, Ercisli, 2015). Use a feedback loop from damage of rate of damage, and express the rate of damage as a function of the current damage level and a fractional rate. The fractional rate is as follows.

$$\text{Fractional rate} = (\text{Maximum fractional rate})(1 - \text{Damage} / (\text{Maximum damage}))$$

where maximum fraction rate = 0.4 (specified) and maximum damage = 1.

Now, once the parameters have been stated, simulate the model, and plot the results; see Figures 7.8–7.17. In Figure 7.8, note the relatively higher accumulation damage/shorter life for the case where the interdependence is considered compared with the case where it is not considered. The surface layer reaches full damage in 16 years in the dependent case compared with 20 years for the independent case (only logistic growth in damage due to traffic), due to the feedback loop.

Figure 7.9 shows that the rate of damage increases and then decreases over time. The rate of damage from the base peaks at the same time as the base damage rate due to the surface damage. In this case, the damage rate of the base has been considered to be impacted fully by the damage rate of the surface in 5 years – a time delay (can be adjusted according to observations) that is expected because of the time needed by the water to reach from the cracked surface to the base.

Figure 7.10 shows that the time to reach a damage of 0.8 years for the surface ranges from 7 to 10 years for a difference in delay from 1 to 8 years; this information is useful to select suitable sealing operations of the surface, based on their effectiveness and cost, compared with the cost of lost service life of the pavement. Figure 7.11 shows the derived equation for the effect of delay on the time for surface damage to reach 0.8.

| k, Uncracked HMA, cm/s | k, Crack, cm/s | % Cracking | k, Cracked HMA, cm/s | k, Cracked HMA, m/s |
|---|---|---|---|---|
| 0.00001 | 0.0206 | 2 | 0.0004218 | 0.000004218 |
| | | 6 | 0.0012454 | 0,000012454 |
| | | 10 | 0.002069 | 0.00002069 |
| | | 15 | 0.0030985 | 0.000030985 |
| | | 20 | 0.004128 | 0.00004128 |

| hHMA, m | hBase, m | keffective, m/s | Effective k, m/s | Saturated_Volumetric_Moisture_Content_% |
|---|---|---|---|---|
| 0.15 | 0.6 | 1.9054E-05 | 1.9054E-05 | 0.31 |
| | | 4.73358E-05 | 4.73358E-05 | 0.31 |
| | | 6.78746E-05 | 6.78746E-05 | 0.31 |
| | | 8.67959E-05 | 8.67959E-05 | 0.31 |
| | | 0.000100893 | 0.000100893 | 0.31 |

| Initial_Volumetric_Moisture_Content | Total_Infiltration_Thickness_m | Depth_of_Ponded_Water_m | Matric_Suction_m |
|---|---|---|---|
| 0.06401 | 0.75 | 0.0254 | 0.1 |
| 0.06401 | 0.75 | 0.0254 | 0.1 |
| 0.06401 | 0.75 | 0.0254 | 0.1 |
| 0.06401 | 0.75 | 0.0254 | 0.1 |
| 0.06401 | 0.75 | 0.0254 | 0.1 |

| Time, s, to Saturation for Base | Time, Hour, to Saturate | 0.5 Hour Rainstorm, Fraction Saturated | Resultant MR, MPa |
|---|---|---|---|
| 6536.74 | 1.81 | 0.27 | 338.14 |
| 2631.22 | 0.73 | 0.68 | 147.34 |
| 1835.01 | 0.50 | 0.98 | 124.34 |
| 1434.98 | 0.39 | 1 | 121.87 |
| 1234.49 | 0.34 | 1 | 121.87 |

| Optimum Mr, MPa | Base Damage |
|---|---|
| 250 | 0 |
| | 0.41 |
| | 0.50 |
| | 0.51 |
| | 0.51 |

| % Surface Cracking | Surface Damage | Base Damage |
|---|---|---|
| 2 | 0.1 | 0.00 |
| 6 | 0.3 | 0.41 |
| 10 | 0.5 | 0.50 |
| 15 | 0.75 | 0.51 |
| 20 | 1 | 0.51 |

| k, uncracked HMA, cm/s | k, crack, cm/s | % Cracking | k, cracked HMA, cm/s | k, cracked HMA, m/s |
|---|---|---|---|---|
| 0.00001 | 0.0206 | 2 | 0.0004218 | 0.000004218 |
| | | 6 | 0.0012454 | 0.000012454 |
| | | 10 | 0.002069 | 0.00002069 |
| | | 15 | 0.0030985 | 0.000030985 |
| | | 20 | 0.004128 | 0.00004128 |

| hHMA, m | hBase, m | keffective, m/s | Effective k, m/s | Saturated_volumetric_moisture_content_% |
|---|---|---|---|---|
| 0.15 | 0.6 | 1.9054E-05 | 1.9054E-05 | 0.31 |
| | | 4.73358E-05 | 4.73358E-05 | 0.31 |
| | | 6.78746E-05 | 6.78746E-05 | 0.31 |
| | | 8.67959E-05 | 8.67959E-05 | 0.31 |
| | | 0.000100893 | 0.000100893 | 0.31 |

| initial_volumetric_moisture_content | Total_infiltration_thickness_m | Depth_of_ponded_water_m | Matric_Suction_m |
|---|---|---|---|
| 0.06401 | 0.75 | 0.0254 | 0.1 |
| 0.06401 | 0.75 | 0.0254 | 0.1 |
| 0.06401 | 0.75 | 0.0254 | 0.1 |
| 0.06401 | 0.75 | 0.0254 | 0.1 |
| 0.06401 | 0.75 | 0.0254 | 0.1 |

| Time, s, to saturation for base | Time, hour, to saturate | 0.5 hour rainstorm, fraction saturated | Resultant MR, MPa |
|---|---|---|---|
| 6536.74 | 1.81 | 0.27 | 338.14 |
| 2631.22 | 0.73 | 0.68 | 147.34 |
| 1835.01 | 0.50 | 0.98 | 124.34 |
| 1434.98 | 0.39 | 1 | 121.87 |
| 1234.49 | 0.34 | 1 | 121.87 |

| Optimum Mr, MPa | Base Damage |
|---|---|
| 250 | 0 |
| | 0.41 |
| | 0.50 |
| | 0.51 |
| | 0.51 |

| % Surface cracking | Surface Damage | Base Damage |
|---|---|---|
| 2 | 0.1 | 0.00 |
| 6 | 0.3 | 0.41 |
| 10 | 0.5 | 0.50 |
| 15 | 0.75 | 0.51 |
| 20 | 1 | 0.51 |

$y = -3.85x^4 + 11.04x^3 - 11.69x^2 + 5.45x - 0.44$

**FIGURE 7.6** Derivation of equation of base damage as a result of surface damage (see from left to right, top to bottom) (continued).

In Figure 7.12, note the difference in design life from 20 to 12 years for unsaturated versus constantly saturated base. Figure 7.13 shows the damage versus time for surface for a flooding event in the fifth year. Flooding causes inundation of the pavement. A 25 mm depth of water on the surface with 6% cracks can reach the entire base in less than 1 hour and keeps the base saturated (Nivedya et al., 2020); the result is a significant increase in damage rate. In Figure 7.14, the damage versus time for surface is shown for a flooding event in the fifth year and subsequent recovery of the base properties to pre-flood levels within 1 year. Note the delay in reaching a damage of 1 (18 years) compared with Figure 7.13 (14 years).

| Base damage | Mr, Base, Mpa | Tensile strain, HMA bottom | Nf, fatigue | Nf, fatigue, design |
|---|---|---|---|---|
| 0.01 | 250 | -0.000197 | 2,655,257 | 2,655,257 |
| 0.25 | 187.5 | -0.00022 | 1,846,213 | |
| 0.33 | 167.5 | -0.000229 | 1,617,992 | |
| 0.5 | 125 | -0.000253 | 1,165,535 | |

| Nf, fatigue / Nf, fatigue, design | % Cracking | Surface Damage, fraction |
|---|---|---|
| 1 | 0.73 | 0.03 |
| 0.69 | 6.62 | 0.33 |
| 0.60 | 8.29 | 0.41 |
| 0.43 | 11.59 | 0.57 |

| Nf, fatigue / Nf, fatigue, design | % Cracking |
|---|---|
| 0.98 | 1 |
| 0.75 | 6 |
| 0.5 | 10 |
| 0.01 | 20 |

**Defined**
20% cracking is 100% surface damage

| Base damage | Surface Damage |
|---|---|
| 0.01 | 0.04 |
| 0.25 | 0.33 |
| 0.33 | 0.41 |
| 0.5 | 0.58 |

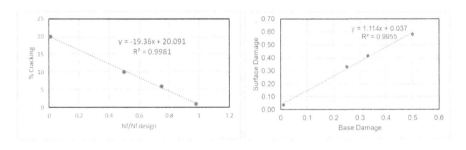

**FIGURE 7.7** Derivation of equation of surface damage as a result of base damage (see from left to right, top to bottom).

Now if we consider the results of this analysis to come up with potential solutions to the problem, we realize that reducing the permeability of the surface asphalt layer is the single most practical step that we can take. This can be achieved to a certain extent by achieving high density and making sure that the surface is crack-free. However, due to spatial and material variabilities, some spots with high permeabilities will be unavoidable. Increasing the thickness may also not be practical due to geometric and economic considerations. One of the time-tested solutions for such a problem is the provision of an asphalt-saturated paving fabric (Marienfeld and Baker, 1999) just beneath the asphalt surface layer, which will reduce the permeability significantly.

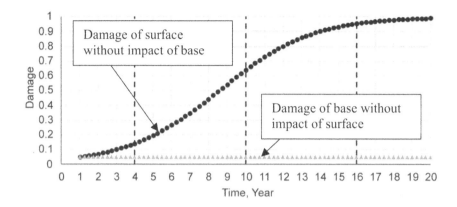

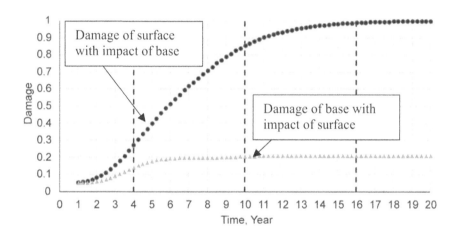

**FIGURE 7.8**   Plots of damage versus time with and without consideration of interdependence of base and surface and cascading effect.

Figure 7.15 shows damage versus time for paving fabrics with different hydraulic conductivities. Note that with a decrease in hydraulic conductivity, which reduces the infiltration of water into the base from the surface, there is a corresponding increase in time over which damage reaches a value of 1. Since there are different types of paving fabrics and the permeability does depend on asphalt saturation, one question is what is a desirable permeability? Figure 7.15 shows that a fabric with 10–1 mm/s permeability does not make any difference, with 10–3 mm/s, the beneficial effect is only for the first few years, whereas with 10–5 mm/s permeability, the effect is the same as that of assuming no contribution of the base toward the surface damage, that is, makes the surface layer fully insulated from the base. This provides a solution to the problem of cascading effect of water infiltration into pavements.

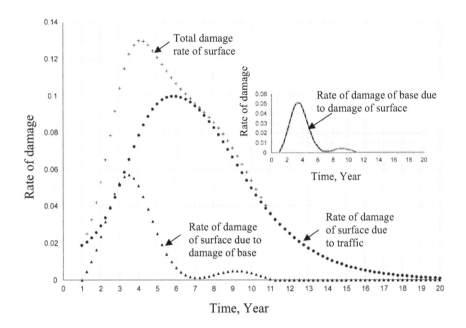

**FIGURE 7.9**  Plots of rate of damage of surface versus time with consideration of rate of damage of base.

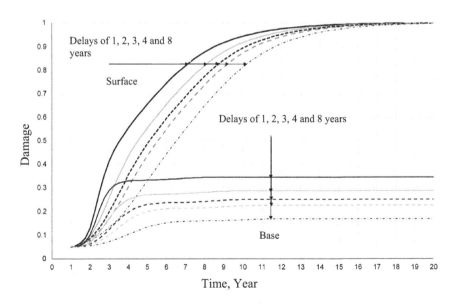

**FIGURE 7.10**  Plots of damage of surface versus time for delays in the impact of the base by surface damage.

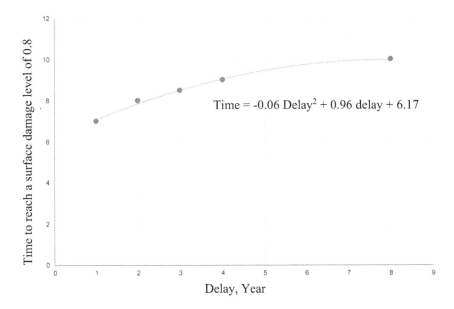

**FIGURE 7.11**   Effect of delay on the time for surface damage to reach 0.8.

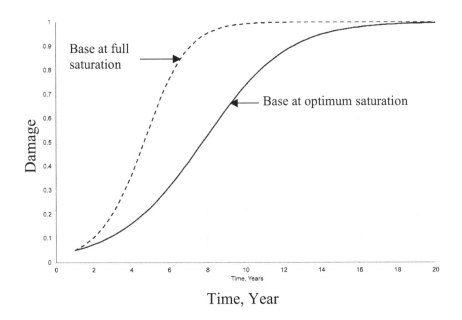

**FIGURE 7.12**   Plot of damage versus time for surface for different base saturation.

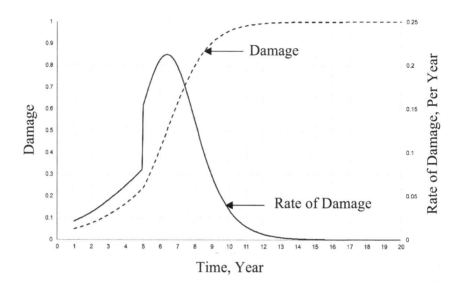

**FIGURE 7.13**   Plot of damage versus time for surface for a flooding event in the fifth year.

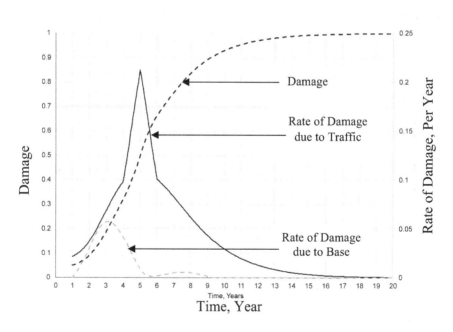

**FIGURE 7.14**   Plot of damage versus time for surface for a flooding event in the fifth year and subsequent recovery of the base properties to pre-flood levels within 1 year.

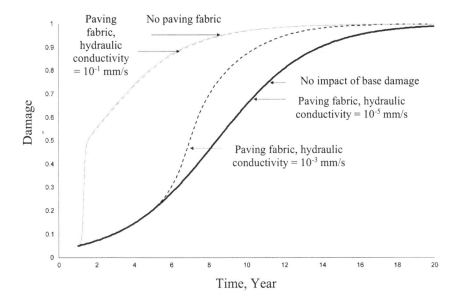

**FIGURE 7.15** Plot of damage versus time for paving fabrics with different hydraulic conductivities.

Now, if it is more feasible to seal cracks and limit the permeability by sealing cracks, rather than providing a fabric (such as for economic or logistic reasons), the question is up to what extent the cracks need to be sealed. Figure 7.16 shows the damage versus time with percentages of crack sealing; as the percentage of sealing increases, the time to reach a damage of 1 increases; there is almost no difference in the time between no sealing and only 25% sealed cracks case; on the other hand, for 90% crack sealing, the loss in life, as calculated by the reduction in time to reach a damage of 1, is only 10%, compared with a case where there is no impact of the base on the surface. The data from Figure 7.16 are used to plot (shown in Figure 7.17) the time to reach a damage of 0.8 (which may be considered as terminal damage when rehabilitation is required) and pavement life loss for different percentages of crack sealing. Note that for 90% sealed cracks, the damage takes 11 years to reach 0.8, compared with 12 years for a pavement with no cracks, and the loss of life is about 8%.

Through the example in this chapter, we have formulated a problem that is not usually considered in pavement analysis, modeled the problem in system dynamics, demonstrated the cascading effects of problems in one/different component/s of a system on the system (pavement) itself, and used the model to evaluate the impact of several potential solutions of the problem.

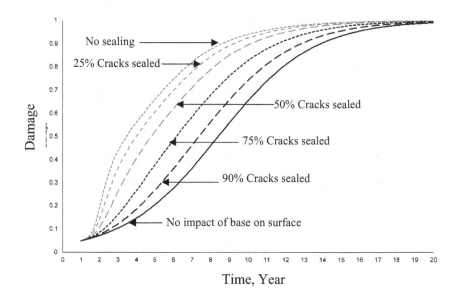

**FIGURE 7.16** Plot of damage versus time for paving fabrics with percentages of crack sealing.

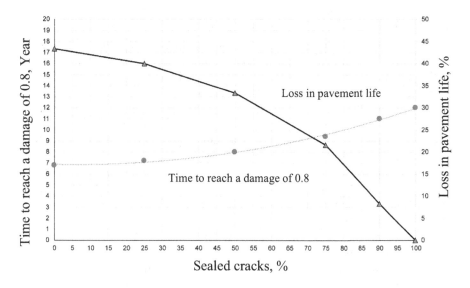

**FIGURE 7.17** Plot of time to reach a damage of 0.8 and pavement life loss for different percentages of crack sealing.

# REFERENCES

ARA. (2004). *Final Report. Part 3. Design Analysis. Chapter 6. HMA Rehabilitation of Existing Pavements. Guide for Mechanistic-Empirical Design of New and Rehabilitated Pavement Structures.* National Cooperative Highway Research Program, Washington, DC.

Brown, E.R., Hainin, M.R., Cooley, A., and Hurley, G. (2008). Relationship of HMA in-place air voids, lift thickness, and permeability. Volume three. NCHRP Web Document 68 (Project 9–27), Transportation Research Board (TRB), Washington, DC.

Ercisli, S. (2015). *Development of Enhanced Pavement Deterioration Curves.* MS thesis, Virginia Polytechnic Institute and State University, Blacksburg, VA.

Green, W.H. and Ampt, G. (1911). Studies of soil physics, part I – the flow of air and water through soils. *The Journal of Agricultural Science*, 4, 1–24.

Haider, S. W., Masud, M. M., and Chatti, K. (2019). Influence of moisture infiltration on flexible pavement cracking and optimum timing for surface seals. *Canadian Journal of Civil Engineering.* doi: 10.1139/cjce-2019-0008.

Main Roads. (2003). Western Australia. Engineering Road Note No. 5.

Mallick, R.B. (2021). A simulative approach to evaluate the feedback effect of water infiltration in pavements. *Infrastructure Asset Management.* doi: 10.1680/jinam.21.00002

Marienfeld, M. L. and Baker, T. L. (1999). Paving Fabric Interlayer as a Pavement Moisture Barrier. Transportation Research Board, Washington, DC, USA, Transportation Research Circular No. E-C006.

Nivedya, M. K., Tao, M., Mallick, R. B., Daniel, J., and Jacobs, J. M. (2020). A framework for the assessment of contribution of base layer performance towards resilience of flexible pavement to flooding. *International Journal of Pavement Engineering*, 21(10), 1223–1234, doi: 10.1080/10298436.2018.1533637.

Ridgeway, H. H. (1976). Infiltration of water through the pavement surface, transportation research board. *Transportation Research Record*, 616, 98–100.

Saeed, K. (2015). *Latent Capacity Support System. System Dynamics SD 557 Class Notes.* Worcester, MA: WPI.

# 8 Analysis of Pavement Systems

Recall that in Chapter 3, we have given several examples of systems that are applicable to pavement engineering. In this chapter, we consider one of those examples, the asphalt mix, as a system, and develop and simulate models to evaluate the change in its critical properties over time. Since different factors of a pavement system are known to interact with each other, the problem is predicting the value of the critical parameter/s over time, by considering all the pertinent factors and their relationships. Furthermore, we consider a specialized mix, porous friction course (PFC), for which design, construction, and regular maintenance are known to play significant roles in defining its performance over time. Hence, here, we illustrate with two examples with the following principle. Principle 4. A pavement system's properties and performance are dictated by the interaction of several factors, which may include design, construction, and maintenance

## EXAMPLE PROBLEM 1: INTERRELATIONSHIP BETWEEN DIFFERENT PROPERTIES OF HOT MIX ASPHALT

1. Define the problem and the boundary of the system
   a. What precisely is the problem?
      The problem is of understanding the change in modulus (considered at a single temperature of 25°C, for simplicity) of hot mix asphalt (or asphalt mix) of in-place pavement over time. This is a complex problem because the modulus is affected by multiple factors that are correlated and have feedback.
   b. What are the factors, interrelationships, and feedback that are known to be associated with the problem, which will constitute the modeled "system"?
      The factors that constitute the system to be modeled are indicated in Table 8.1.
   c. What is the appropriate time scale of the problem?
      Let us consider a time of 20 years.
   d. What are the known behaviors of the relevant dynamic factors (from existing data/literature)?
      i. Voids in total mix (VTM), %: Known to decrease nonlinearly over time; the rate of change slows down over time.
      ii. Asphalt binder viscosity (As), Poise: Known to increase nonlinearly over time, the rate of change slows down over time.

DOI: 10.1201/9781003345596-8

## TABLE 8.1
### Relevant Factors

| Factors That Change Over Time or Can Vary | Factors That Are Considered to Remain Constant |
| --- | --- |
| Voids in total mix (VTM), % | Percent binder in the mix (Pb) |
| Voids in mineral aggregate (VMA), % | Bulk specific gravity of aggregate ($G_{sb}$) |
| Voids filled with asphalt (VFA), % | Volume of bulk aggregate ($V_{sb}$) |
| Asphalt binder viscosity (As), Poise | Initial VTM, % |
| Dynamic modulus E at 25°C | Volume of effective asphalt binder ($V_{be}$) |
| Frequency of loading | Volume associated with mix, $V_{mm}$ |
| | Maximum specific gravity of mix, $G_{mm}$ |
| | Initial bulk specific gravity of mix, $G_{mb}$ |
| | Cumulative retained on 4.75 mm sieve |
| | Cumulative retained on 9.5 mm sieve |
| | Cumulative retained on 19 mm sieve |
| | Percent passing 75 micron sieve |

    iii. Dynamic modulus (modulus), psi: can be expressed as a function of VTM, As, volume of effective asphalt binder, frequency of loading, cumulative retained on 19, 9.5 and 4.75 mm sieves, percent passing the 75 micron sieve.

2. Develop a dynamic hypothesis of the problem

    a. Tie in the known causes and theories related to the problem

        i. The mix has an initial modulus. Over time, it changes because of changes in VTM and As. For all other factors remaining constant, the change in modulus can be explained by the changes in VTM and As.

        ii. The change in VTM is nonlinear over time; hence, the change must be a function of both time and VTM at any time.

        iii. The change in As is nonlinear over time; hence, the change must be a function of both time and As at any time.

    b. Develop a causal structure that explains the behavior of the system in terms of relevant factors, their changes over time, and feedback:

        i. Change in VTM = f(VTM, Time)
= (VTM Coeff*VTM) + (VTM Time Coeff*TIME) + (VTM Time2 Coeff*TIME^2);
Where VTM Time2Coeff is a negative number

        ii. Change in asphalt binder viscosity = f(Viscosity, TIME)
= (Viscosity Coefficient*Viscosity) + (Vis Time Coefficient* TIME) + (Vis Time2 Coefficient*TIME^2)
Where Vis Time2 Coefficient is a negative number

    c. Draw the causal structure

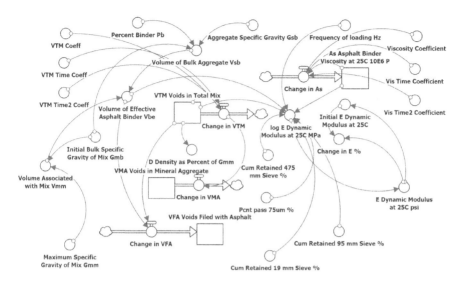

**FIGURE 8.1** Causal structure.

The causal structure is shown in Figure 8.1. Here, STELLA v10.0.4 has been used to simulate the model for a time span of 20 years. During simulation, the software computes the value of each parameter at small time steps – in this case, a time step of 0.25 year was used. The parameters are shown in Table 8.2.

Assign appropriate equations and values. The appropriate equations and values are shown in Table 8.2. Next, simulate the model. See the results in Figure 8.2. The plots match well with existing observation/data (Brown et al., 2016).

## A MODEL TO DETERMINE THE PERFORMANCE OF POROUS FRICTION COURSE

PFCs can be used effectively to reduce wet weather-related hazards such as poor visibility and hydroplaning (Cooley et al., 2009). The advantages of these mixes are that they allow quick draining of water and provide higher friction; they can also be used for reducing noise. Such mixes are, however, also prone to early onset of distress and progressive clogging of voids by winter treatment materials, unless they are designed properly, and appropriate maintenance (vacuum cleaning) practices are adopted.

Therefore, for evaluation of PFC, a system of multi-criterion simulation and analysis for a holistic evaluation in different climatic conditions is needed. The objectives of the model presented in this chapter are to develop a framework for a system for PFCs and evaluate the long-term performance of porous graded asphalt using a multi-criterion analysis approach. This model was created with several key interrelated factors and their relationships, as well as the effect of environment on them.

The variables associated with the materials and design include the percent binder content, the fiber dosage, and the polymer content (both of which can allow the use of higher binder contents and hence better durability). Site-specific factors include MAAT (Mean Annual Air Temperature) and proximity to the construction site, both

## TABLE 8.2
## Equations and Values

Top-Level Model:

As_Asphalt_Binder_Viscosity_at_25°C_10E6_P(t) = As_Asphalt_Binder_Viscosity_
at_25°C_10E6_P(t − dt) + (Change_in_As) * dt

INIT As_Asphalt_Binder_Viscosity_at_25°C_10E6_P = 3,000

INFLOWS:

Change_in_As = (Viscosity_Coefficient/
As_Asphalt_Binder_Viscosity_at_25°C_10E6_P) + (Vis_Time_Coefficient*TIME) + (Vis_Time2_
Coefficient*TIME^2)

VFA_Voids_Filled_with Asphalt(t) = VFA_Voids_Filled_with_Asphalt(t − dt) + (Change_in_VFA) * dt

INIT VFA_Voids_Filled_with_Asphalt = 53.94

INFLOWS:

Change_in_VFA = 100* (Volume_of_Effective_Asphalt_Binder_Vbe)/(Volume_of_Effective_Asphalt_
Binder_Vbe + VTM_Voids_in_Total_Mix) − 100* (INIT(Volume_of_Effective_Asphalt_Binder_Vbe))/
(INIT(Volume_of_Effective_Asphalt_Binder_Vbe)+INIT(VTM_Voids_in_Total_Mix))

VMA_Voids_in_Mineral_Aggregate(t) = VMA_Voids_in_Mineral_Aggregate(t − dt) + ( −Change_in_
VMA) * dt

INIT VMA_Voids_in_Mineral_Aggregate = 16.73

OUTFLOWS:

Change_in_VMA = Change_in_VTM

VTM_Voids_in_Total_Mix(t) = VTM_Voids_in_Total_Mix(t − dt) + (−Change_in_VTM) * dt

INIT VTM_Voids_in_Total_Mix = 8

OUTFLOWS:

Change_in_VTM = (VTM_Coeff*VTM_Voids_in_Total_Mix)+(VTM_Time_Coeff*TIME)+(VTM_
Time2_Coeff*TIME^2)

Aggregate_Specific_Gravity_Gsb = 2.65

Change_in_E_% = 100*(E_Dynamic_Modulus_at_25°C_psi-Initial_E_Dynamic_Modulus_at_25C)/
Initial_E_Dynamic_Modulus_at_25°C

Cum_Retained_19_mm_Sieve_% = 0

Cum_Retained_475_mm_Sieve_% = 37

Cum_Retained_95_mm_Sieve_% = 3

D_Density_as_Percent_of_Gmm = 100-VTM_Voids_in_Total_Mix

E_Dynamic_Modulus_at_25°C_Mpa = E_Dynamic_Modulus_at_25°C_psi*6.9/1,000

E_Dynamic_Modulus_at_25°C_psi = 10^log_E_Dynamic_Modulus_at_25°C_MPa

Frequency_of_loading_Hz = 10

Initial_Bulk_Specific_Gravity_of_Mix_Gmb = 2.335

Initial_E_Dynamic_Modulus_at_25°C = INIT(E_Dynamic_Modulus_at_25°C_psi)

log_E_Dynamic_Modulus_at_25°C_MPa = 3.750063 + 0.02932*Pcnt_pass_75um_%-0.001767*(Pcnt_
pass_75um_%^2) − 0.002841*Cum_Retained_475_mm_Sieve_%-0.058097*VTM_Voids_in_Total_
Mix-0.0802208*(Volume_of_Effective_Asphalt_Binder_Vbe/
(Volume_of_Effective_Asphalt_Binder_Vbe + VTM_Voids_in_Total_
Mix)) + ((3.871977 − 0.0021*Cum_Retained_475_mm_Sieve_% + 0.003958*Cum_Retained_95_mm_
Sieve_%-0.000017*(Cum_Retained_95_mm_Sieve_%^2) + 0.005470*Cum_Retained_19_mm_
Sieve_%)/
(1 + EXP(−0.603313 − 0.313351*LOG10(Frequency_of_loading_Hz)-0.393532*LOG10(As_Asphalt_
Binder_Viscosity_at_25°C_10E6_P))))

*(Continued)*

**TABLE 8.2** (*Continued*)
**Equations and Values**

Maximum_Specific_Gravity_of_Mix_Gmm = 2.53
Pcnt_pass_75um_% = 6
Percent_Binder_Pb = 5.5
Vis_Time_Coefficient = 0.1
Vis_Time2_Coefficient = −0.5
Viscosity_Coefficient = 1,000,000
Volume_Associated_with_Mix_Vmm = Initial_Bulk_Specific_Gravity_of_Mix_Gmb/
  Maximum_Specific_Gravity_of_Mix_Gmm
Volume_of_Bulk_Aggregate_Vsb = (Initial_Bulk_Specific_Gravity_of_Mix_Gmb-Percent_Binder_
  Pb*Initial_Bulk_Specific_Gravity_of_Mix_Gmb/100)/Aggregate_Specific_Gravity_Gsb
Volume_of_Effective_Asphalt_Binder_Vbe = Volume_Associated_with_Mix_Vmm-Volume_of_Bulk_
  Aggregate_Vsb
VTM_Coeff = 0.1
VTM_Time_Coeff = 0.01
VTM_Time2_Coeff = −0.001
{ The model has 32 (32) variables (array expansion in parens).
In root model and 0 additional modules with 0 sectors.
Stocks: 4 (4) Flows: 4 (4) Converters: 24 (24)
Constants: 15 (15) Equations: 13 (13) Graphicals: 0 (0)
}

of which are linked to clogging potential because depending on the MAAT, sand/ ice/de-icing materials that tend to clog PFCs may or may not be used, and traffic can bring clogging materials from the construction site.

The clogging is related to the decrease of voids in total mix (voids) in the PFC as well as to the annual maintenance (vacuum) frequency. The durability is related to the stiffening of the binder, represented by changes in the viscosity, which is related to the MAAT. The change in permeability was related to the change in air voids since it has been shown to affect the permeability of asphalt mixes significantly. The change in noise absorption has been related to permeability, while the friction number has been related to the polished value of the aggregates. Finally, change in rutting has been related to the polymer content, since the use of polymer has been shown to be very effective in reducing the rutting potential of asphalt mixes. Finally, weightages were given to the different distress potentials, and a combined durability index was determined over time.

The causal structure is shown in Figure 8.3. The key variables are shown in Table 8.3. Note that the values of the parameters (represented by converters) are taken for the base case scenario in this table. For example, the maintenance frequency (N per year) is considered to be 0.

Now let us simulate the model. Since the site is close to a construction area (proximity is set at 1, Table 8.3), it is not surprising to see in the results (Figure 8.4) that the permeability drops rapidly over time, as the voids get clogged over time with debris/ particles. Now, one can experiment with the model and observe the changes in the

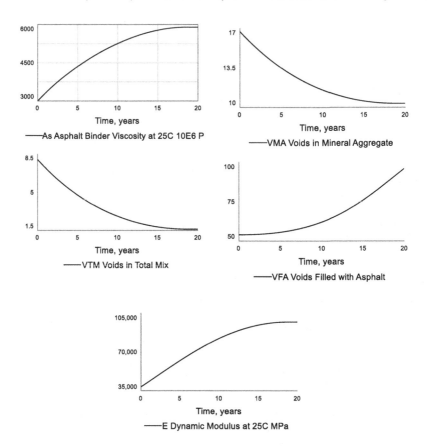

**FIGURE 8.2** Plots of hot mix asphalt properties versus time.

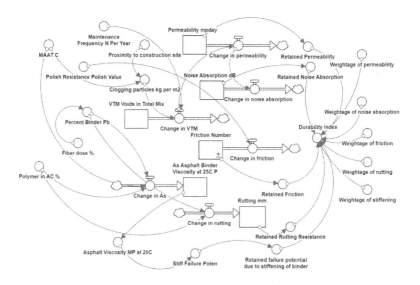

**FIGURE 8.3** System dynamics model of porous friction course.

## TABLE 8.3
## Key Variables

Top-Level Model:

As_Asphalt_Binder_Viscosity_at_25°C_P(t) = As_Asphalt_Binder_Viscosity_

at_25°C_P(t − dt) + (Change_in_As) * dt

INIT As_Asphalt_Binder_Viscosity_at_25°C_P = 2,851,877

INFLOWS:

Change_in_As = (−0.5*10^6*TIME+2*10^7) − 10^6*(Percent_Binder_Pb+5)-5*10^5*Polymer_in_

AC_% + 10^5*MAAT_C

Friction_Number(t) = Friction_Number(t − dt) + (−Change_in_friction) * dt

INIT Friction_Number = 55

OUTFLOWS:

Change_in_friction = 30-TIME*(0.5*Polish_Resistance_Polish_Value)

Noise_Absorption_dB(t) = Noise_Absorption_dB(t − dt) + (−Change_in_noise_absorption) * dt

INIT Noise_Absorption_dB = 3

OUTFLOWS:

Change_in_noise_absorption = IF(Change_in_permeability > 0)THEN(0.025*Change_in_permeability)

ELSE(0)

Permeability_mpday(t) = Permeability_mpday(t − dt) + (−Change_in_permeability) * dt

INIT Permeability_mpday = 120

OUTFLOWS:

Change_in_permeability = IF(Change_in_VTM>0)THEN(2*TIME*Change_in_VTM)ELSE(0)

Rutting_mm(t) = Rutting_mm(t − dt) + (Change_in_rutting) * dt

INIT Rutting_mm = 0.05

INFLOWS:

Change_in_rutting = (0.1 + 0.1*Percent_Binder_Pb)*TIME-2*Polymer_in_AC_%

VTM_Voids_in_Total_Mix(t) = VTM_Voids_in_Total_Mix(t − dt) + (−Change_in_VTM) * dt

INIT VTM_Voids_in_Total_Mix = 18

OUTFLOWS:

Change_in_VTM = IF(Clogging_particles_kg_per_m$^2$ > 0)THEN (−2*Maintenance_Frequency_N_Per_

Year+2*Clogging_particles_kg_per_m$^2$) ELSE(0)

Asphalt_Viscosity_MP_at_25°C = As_Asphalt_Binder_Viscosity_at_25C_P/1,000,000

Clogging_particles_kg_per_m$^2$ = IF MAAT_C > 15.55 THEN (0.05*Proximity_to_construction_site)

ELSE (0.05*Proximity_to_construction_site + 1.65 − 0.09*MAAT_C)

Durability_Index = Weightage_of_permeability*Retained_Permeability +

Weightage_of_friction*Retained_Friction +

Weightage_of_noise_absorption*Retained_Noise_Absorption +

Weightage_of_stiffening*Retained_failure_potential_due_to_stiffening_of_binder +

Weightage_of_rutting*Retained_Rutting_Resistance

Fiber_dose_% = 0.1

MAAT_C = 7

Maintenance_Frequency_N_Per_Year = 0

Percent_Binder_Pb = 5.5 + 1.5*Fiber_dose_%

Polish_Resistance_Polish_Value = 60

Polymer_in_AC_% = 1

Proximity_to_construction_site = 1

(*Continued*)

---

**TABLE 8.3** (*Continued*)

**Key Variables**

Retained_failure_potential_due_to_stiffening_of_binder = INIT(Stiff_Failure_Poten)/Stiff_Failure_Poten

Retained_Friction = Friction_Number/INIT(Friction_Number)

Retained_Noise_Absorption = Noise_Absorption_dB/INIT(Noise_Absorption_dB)

Retained_Permeability = Permeability_mpday/INIT(Permeability_mpday)

Retained_Rutting_Resistance = INIT(Rutting_mm)/Rutting_mm

Stiff_Failure_Poten = IF(Asphalt_Viscosity_MP_at_25°C < 71)THEN(1)ELSE(2)

Weightage_of_friction = 0.15

Weightage_of_noise_absorption = 0.1

Weightage_of_permeability = 0.4

Weightage_of_rutting = 0.05

Weightage_of_stiffening = 0.3

{ The model has 33 (33) variables (array expansion in parens).

In root model and 0 additional modules with 0 sectors.

Stocks: 6 (6) Flows: 6 (6) Converters: 21 (21)

Constants: 11 (11) Equations: 16 (16) Graphicals: 0 (0)

}

---

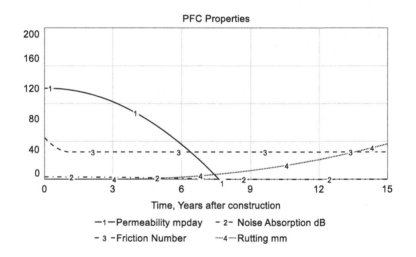

**FIGURE 8.4**    Plots of porous friction course properties versus time: Base case scenario.

properties with a change in the controllable factors. For example, if maintenance, in the form of vacuum cleaning of debris is conducted three times a year, will it make a difference? See Figure 8.5, which shows that permeability and associated properties such as noise absorption and friction number remain unchanged over the years as a result of regular maintenance. However, rutting still increases as before because it is a function of (in addition to other factors) the polymer content in the asphalt binder, which is still set to a very low content of 1% in the current run. Figure 8.6 shows the effect of a higher polymer content (2.5%) – the rate of increase in rutting

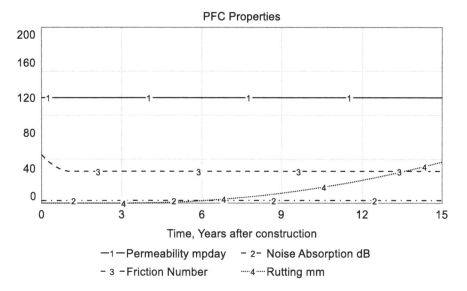

**FIGURE 8.5** Plots of porous friction course properties versus time: Case with regular maintenance.

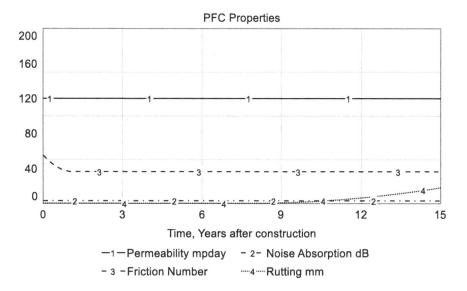

**FIGURE 8.6** Plots of porous friction course properties versus time: Case with regular maintenance and higher polymer content.

is significantly reduced – it reaches a value of 18 mm in 15 years (which may be the design life of the layer). However, we must note here that the trends in the data are more important than the actual numbers because the numbers do depend on the calibration of the model with existing data.

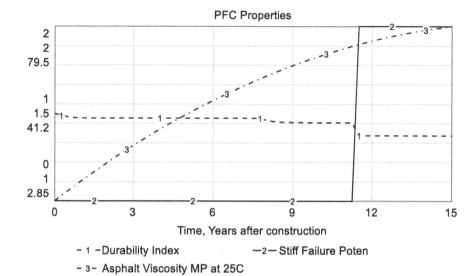

**FIGURE 8.7**   Plots of porous friction course properties versus time: Durability index, stiffening failure potential, and asphalt viscosity.

One more important property deserves attention. Since PFC mixes have high air voids (initial and constantly, if maintained properly), there is an enhanced potential of faster aging and hence stiffening of the binder in the entire layer (remember, these layers are relatively thin, typically 25–50 mm thick), and resultant cracking. Such stiffening, in addition to voids, can be modeled as functions of asphalt binder and polymer content. An increase in both of these can reduce the potential of stiffening and cracking. In our model, we have defined a durability index that takes into consideration this cracking potential (see Table 8.3) in addition to the properties listed above. The cracking potential is set to increase when the asphalt viscosity exceeds a critical value. As Figure 8.7 shows, when the asphalt viscosity exceeds this critical value, the stiffening potential increases, and the durability index drops. The list of above properties is, by no means, exhaustive; one may include other properties as desired, in the same way.

## REFERENCES

Brown, E. R., Kandhal, P. S., Roberts, F. L., Kim, Y. R., and Lee, D.-Y. (2016). *Hot Mix Asphalt Materials, Mixture Design and Construction*, 3rd Edition, Greenbelt, MD: National Asphalt Pavement Association (NAPA).

Cooley, L. A., Jr., J. W. Brumfield, R. B. Mallick, W. S. Mogawer, M. N. Partl, L. D. Poulikakos, and G. Hicks. (2009). *NCHRP Report 640: Construction and Maintenance Practices for Permeable Friction Courses*. Washington, DC: Transportation Research Board of the National Academies.

# 9 Multidisciplinary Analysis

In this chapter, the following two principles are presented with example models. The first one lays down the foundation for considering both economics and the environment in pavement engineering, while the second one proposes a line of reasoning for considering a limit to growth for pavement networks. The key factors in the two principles are **multidisciplinary consideration** and a **realistic/rational** view of pavements with respect to our world, respectively. Principle 5: Economics and the environment constitute **two critical components** of a sustainable pavement system. Principle 6: System dynamics imposes a **limit to growth** in terms of pavement system size to maintain a reasonable quality of the system.

## COMPOSITE INDEX

Consider the concept of sustainability, which has been defined in many ways by many authors and organizations. Let us attempt to define sustainability as a concept in the realm of road building, maintenance, and rehabilitation. We all know that poor-quality roads (engineering) lead to a loss of quality of life, higher costs of road construction (primarily due to increase in fuel cost), erosion of our budget (economics), and emission of $CO_2$ during construction leads to a deterioration of the environment. Therefore, a measure of Sustainability Index over time can be made by considering the combined impact of changes in road quality, fuel cost, and $CO_2$ emission. Conceptually, this can be represented as in Figure 9.1.

Note that in this case, the Sustainability Index is a **Composite Index**, which allows us to evaluate the combined effect of multiple parameters. Such indices are very useful for systems that involve performances or results that are related to multiple criteria.

Now let us try to represent the above relationships mathematically, by considering each parameter separately. First, consider carbon dioxide. Let us say that initially, we have $CO_2$ in the system, and over time, the total $CO_2$ in the system changes as a result of more $CO_2$ being generated by road work. Hence, one way of representing the parameter related to $CO_2$ mathematically is through a ratio:

$$CO_2 \text{ ratio}(t) = \text{Total } CO_2 \text{ in the system}(t) / \text{Initial } CO_2 \text{ in the system}$$

Similarly, the contribution from fuel cost can be indicated as

$$\text{Fuel cost ratio }(t) = \frac{\text{Cost of fuel spent per km of road construction }(t)}{\text{Initial cost of fuel spent per km of road construction}}$$

Finally, let us consider the road quality parameter. Roads are of two types – new roads and those that are ready for rehabilitation. Kilometers of new roads that are needed depend on the growth of a community/city or country. Kilometers of roads that are rehabilitated depend on the average life of the roads and the available budget.

DOI: 10.1201/9781003345596-9

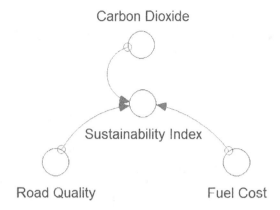

**FIGURE 9.1**  Example of consideration of combined impact.

Altogether, the road quality is a combined function of the following questions: (a) how many new kilometers of roads are built versus how many are needed and (b) how many kilometers of roads are ready for rehabilitation and how many are actually rehabilitated?

Hence, we can write that

$$\text{Road Quality} = (\text{Initial\_road\_quality} * ((1 - \text{New\_road\_shortfall})^{\wedge}0.5)$$

$$* ((1 - \text{Rehabilitation\_shortfall})^{\wedge}0.5))$$

where

New_road_shortfall = (Required_new_roads_per_year-Paving_of_new_roads)/Required_new_roads_per_year

Rehabilitation_shortfall = (Indicated_paving_for_rehabilitation-Paving_for_rehabilitation)/Indicated_paving_for_rehabilitation

Therefore, the relationships can now be mapped as in Figure 9.2.

Suppose an agency gives more weightage to the economic consideration than the engineering consideration. To provide that flexibility, another set of factors need to be added to the map, the weightage factor for each consideration.

Mathematically, we now can write the equation for Sustainability Index as follows:

$$\text{Sustainability Index} = (\text{Road Quality}^{\wedge}\text{Weight Given to Engineering})$$

$$* ((1/\text{Carbon Dioxide Ratio}^{\wedge}\text{Weight Given to Environment}))$$

$$* ((1/\text{Fuel Cost Ratio}^{\wedge}\text{Weight Given to Economics}))$$

Now suppose we want to normalize the Sustainability Index to an initial Index and observe the change over time, then we can indicate a sustainability score as follows:

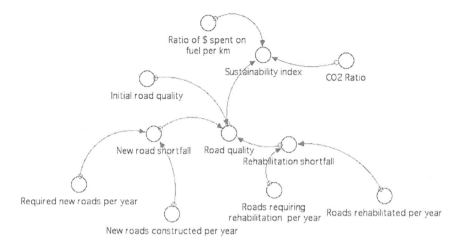

**FIGURE 9.2** Map of the interrelationships.

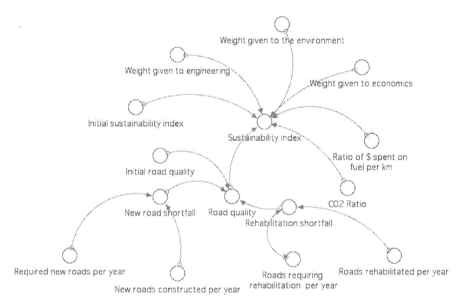

**FIGURE 9.3** Revised map of interrelationships.

$$\text{Sustainability Score} = 100 * \frac{(\text{Sustainability Index}(t))}{(\text{Initial Sustainability Index})}$$

Although all the parameters are with the same symbol (circle) in the last figure, note that some of them are calculated from the model (through simulation over time), while some are assigned. The parameters, for which a value can be assigned, are the converters (e.g., weightage given to economics). The updated map is shown in Figure 9.3.

## COMPLEX PROBLEM – DEFINING THE KEY SECTORS AND PARAMETERS

In the real world, most problems consist of multiple stocks, flows, and converters and interrelationships that are dependent on time. In other words, there would be multiple systems like the one presented in the last chapter, and the flow rates may not be constant. The best approach to model such a complex problem is to map out the different sectors first, write down the relevant parameters (stocks, flows, and converters) within each sector, represent the relationships, provide the mathematical formula (as best as possible in the first cut), run the simulation, check the output against the Reference mode (baseline mode for which data is available from the literature and/ or experiments/experience), and then go back and either include more appropriate parameters and/or improve the mathematical formulae, and then run the simulation again to compare the output against the Reference mode. So modeling is an iterative process, and in every step, you try to improve the model. Remember that the model will never be perfect in the first cut or even after many iterations – but each iteration will generate a "better" model and when we have a workable model, we can start using it to answer our questions.

Let us consider the following problem. We build roads because there is economic/ population growth and demand for new roads. Old roads require maintenance and finally rehabilitation. Road construction activities (new construction, maintenance, and rehabilitation) require aggregates (stones), which may come from a local source or non-local source (which are further away than local sources). If the roads are recycled during maintenance and rehabilitation, then the demand for new aggregates could be reduced. Transporting aggregates to job sites require trips by trucks, which consume diesel, whose cost increases over time due to inflation. In the process, $CO_2$ is generated and added to the atmosphere. $CO_2$ is removed by land, and as the area of forest is reduced (as they are cut down for building new roads), the absorption ability of $CO_2$ changes over time. If new roads are not built and old roads are not maintained or rehabilitated, then the quality of roads suffers, and there is a fall in the engineering quality. Rising fuel costs lead to a deterioration in the economic factor, and finally, a rise in the $CO_2$ in the atmosphere leads to a fall in the environmental quality. How can we model this entire problem and determine a sustainability score of the system at any time? Our intent is to evaluate the effects of different factors on the sustainability score and determine the leverage parameters that can help us maintain sustainable road construction activities.

## STEP 1: IDENTIFY THE KEY SECTORS

Before drawing the different sectors, each of which consists of several stocks, flows, and converters, think about what a 10,000 m view of the problem would be. This part, in fact, is the most important step in the modeling process. From your knowledge, experience, and literature review, what do you think are the important topics in this problem? Ask yourself the question "what are the likely causes and what are the effects"? Let us attempt to write them down, as shown in Table 9.1. Remember that this is not an exhaustive list, but just enough to get going with a model.

## TABLE 9.1
## Important Sectors

| Question | Probable Answers | Identified Sectors |
|---|---|---|
| 1. Why do we need to construct new roads? | As the population grows, the number of vehicles increases; hence, growth in population leads to a demand for new roads | Population and road demand |
| 2. What happens to new roads over time? | Roads deteriorate due to traffic and the environment; roads have a life, say 10 years, after which they need to be rehabilitated; rehabilitation also required construction | Road construction and rehabilitation |
| 3. What is the impact of road construction on the economy? | Road construction requires raw materials, energy, and labor; hence, there are many costs associated with road construction. One important cost is the cost of fuel that is involved in the transportation of aggregates during construction; as the cost of fuel increases, the cost of fuel used for road construction also increases | Economy (consider fuel used during transportation of aggregates only) |
| 4. How does road construction affect the environment? | In many ways. If we consider the transportation of aggregates, the fuel burnt during transportation leads to the emission of $CO_2$; this $CO_2$ is dissipated/absorbed by the trees; if trees are cut down, that is the area of forest is reduced due to road or non-road-related work, then the absorption rate also decreases | Environment sector (consider the generation and absorption of $CO_2$ only) |
| 5. Given the above factors, how can we determine the sustainability of road construction? | By combining the effects of road quality, $CO_2$ generation and dissipation, and fuel costs, we can consider a combined effect of engineering (or quality of transportation or life), the environment, and economy | Sustainability score |

Next, for each sector listed in Table 9.1, let us try to map out the different parameters, as we had done in the last chapter. Let us also try to identify (see in parenthesis) which ones are stocks, which ones are flows, and which ones are converters.

## POPULATION AND ROAD DEMAND SECTOR

Figure 9.4 shows a map of this sector.

Let us now write down the relationships between the parameters, as in Table 9.2. The comments are written in bold and italics. The specific values of converters are assumed or based on literature.

## ROAD CONSTRUCTION AND REHABILITATION SECTOR

First, let us map out the basic parts of this sector, as in Figure 9.5.

The equations relating to the different parameters of this sector can be written as in Table 9.3. You will note that one or more of the parameters have already been defined in the population and demand sector.

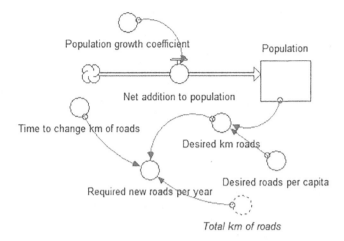

**FIGURE 9.4**    Population and road demand sector.

## TABLE 9.2
### Equations of Parameters in Population and Demand Sector

Population(t) = Population(t − dt) + (Net_addition_to_population) * dt

INIT Population = 1,000,000 (*consider an initial population of 1,000,000*)

Inflows:

Net_addition_to_population = Population_growth_coefficient*TIME

Desired_km_roads = Population*Desired_roads_per_capita

Desired_roads_per_capita, km = 0.009 (*consider 0.009*)

Population_growth_coefficien, people/(person*year) = 170 (*consider 170*)

Required_new_roads_per_year, km = (Desired_km_roads − Total_km_of_roads)

/ (Time_to_change_km_of_roads)

Time_to_change_km_of_roads = 10 years

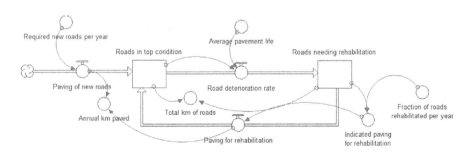

**FIGURE 9.5**    Preliminary road construction and rehabilitation sector.

---

**TABLE 9.3**

**Parameters in the Preliminary Road Construction and Rehabilitation Sector**

Roads_in_top_condition(t), km = Roads_in_top_condition(t − dt) + (Paving_of_new_roads + Paving_
for_rehabilitation − Road_deterioration_rate) * dt

INIT Roads_in_top_condition, km = 8,000 (*consider 8,000*)

*Inflows:*

Paving_of_new_roads, km per year = Required_new_roads_per_year

Paving_for_rehabilitation, km per year = Indicated_paving_for_rehabilitation

*Outflows:*

Road_deterioration_rate, k per year = (Roads_in_top_condition)/(Average_pavement_life)

Roads_needing_rehabilitation(t), km = Roads_needing_rehabilitation(t − dt) + (Road_deterioration_rate
- Paving_for_rehabilitation) * dt

INIT Roads_needing_rehabilitation, km = 100 (consider an initial value of 100 km)

*Inflows:*

Road_deterioration_rate = Roads_in_top_condition/Average_pavement_life

*Outflows:*

Paving_for_rehabilitation, km per year = Indicated_paving_for_rehabilitation

Annual_km_paved = Paving_for_rehabilitation+Paving_of_new_roads

Average_pavement_life, year = 10 (*consider a value of 10*)

Fraction_of_roads_rehabilitated_per_year = 0.7 (*consider a value of 0.7*)

Indicated_paving_for_rehabilitation = Roads_needing_rehabilitation *
Fraction_of_roads_rehabilitated_per_year

Required_new_roads_per_year = (Desired_km_roads -Total_km_of_roads) /
(Time_to_change_km_of_roads)

Total_km_of_roads = Roads_in_top_condition + Roads_needing_rehabilitation

---

Next, consider the above map (and parameters) carefully. Even though it shows the basis of construction, it does not show us the parameters that are important for evaluating the economic and environmental impacts and the consideration of a sustainability score. So what else is needed in this sector? Let us consider them one by one.

1. If we need to consider the impact of transportation of aggregate, we need to know what is the distance traveled for transporting aggregates per year; now in this regard, we can consider two options – transporting aggregates from natural aggregate stocks (e.g., quarries) and after milling from existing roads to plants for recycling aggregates (which can only be done for rehabilitation and not for new road construction). The distance for these two cases will be different. For aggregate transportation from natural aggregate stocks, let us consider the fact that as more and more stock is exhausted, the transportation distance increases (this is evident from the literature); on the other hand, for recycling, let us consider an average distance between the job site (that is the road to be rehabilitated) and the plant (where the aggregate will be recycled in producing new mixes for rehabilitation). Note that the more aggregates we recycle, the less new aggregates we use from the natural aggregate stock.

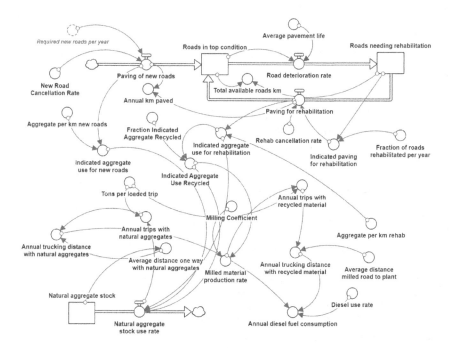

**FIGURE 9.6**   Modified road construction and rehabilitation sector.

2. When we mill the aggregates from existing roads from rehabilitation, we may end up with a fraction (say 95%) of the milled aggregates, and not all of it.
3. The total distance traveled for construction of new roads and rehabilitating roads (by recycling or by using new natural aggregate) needs to be considered for the calculation of the total consumption of fuel. This involves the consideration of total aggregates required and tons per trip.
4. To cope with limited budgets or to reduce the impact on the budget and/or the environment, we may have a policy to rehabilitate only a fraction (say 70%) of roads that need rehabilitation every year and, then furthermore, we may choose to cancel some projects, that is kilometers of new roads and/or kilometers of rehabilitation every year.

Including all the above considerations, we get a modified map and set of equations, as shown in Figure 9.6 and Table 9.4, respectively.

Next, let us consider the economy sector. We need to determine the cost of fuel used for transportation. Since we have already determined the amount of fuel consumed, in the road construction and rehabilitation sector, we need to consider the unit cost of fuel in the sector. However, over time, the cost of fuel increases due to inflation, and we need to include an inflation factor. Considering all of the above, a good way of expressing the cost of fuel is the cost of fuel per kilometer of road paved (we call it a ratio). Therefore, the sector map and the equations can be expressed as shown in Figure 9.7 and Table 9.5, respectively.

## TABLE 9.4
## Updated Set of Parameters for Modified Road Construction and Rehabilitation Sector

Natural_aggregate_stock(t) = Natural_aggregate_stock(t − dt) + (−Natural_aggregate_stock_use_rate) * dt
INIT Natural_aggregate_stock = 1000000000 (**tonnes**)
*Outflows*:
Natural_aggregate_stock_use_rate = Indicated_aggregate_use_for_new_roads+Indicated_aggregate_use_for_
  rehabilitation-Indicated_Aggregate_Use_Recycled
Roads_in_top_condition(t) = Roads_in_top_condition(t − dt) + (Paving_of_new_roads + Paving_for_
  rehabilitation - Road_deterioration_rate) * dt
INIT Roads_in_top_condition = 8000 (**km**)
*Inflows*:
Paving_of_new_roads = Required_new_roads_per_year-New_Road__Cancellation_Rate
Paving_for_rehabilitation = Indicated_paving_for_rehabilitation-Rehab_cancellation_rate
*Outflows*:
Road_deterioration_rate = Roads_in_top_condition/Average_pavement_life
Roads_needing_rehabilitation(t) = Roads_needing_rehabilitation(t − dt) + (Road_deterioration_rate − Paving_
  for_rehabilitation) * dt
INIT Roads_needing_rehabilitation = 100
*Inflows*:
Road_deterioration_rate = Roads_in_top_condition/Average_pavement_life
*Outflows*:
Paving_for_rehabilitation = Indicated_paving_for_rehabilitation-Rehab_cancellation_rate
Aggregate_use_per_km_new_roads = 550 (**tonnes**)
Aggregate_use_per_km_rehabilitated = 300 (**tonnes**)
Annual_diesel_fuel_consumption = (Annual_trucking_distance_with_recycled_material+Annual_trucking_
  distance_with_natural_aggregates)*Diesel_use_rate
Annual_km_paved = Paving_for_rehabilitation+Paving_of_new_roads
Annual_trips_with_natural_aggregates = (Natural_aggregate_stock_use_rate)/Tons_per_loaded_trip
Annual_trips_with_recycled_material = Milled_material__production_rate/Tons_per_loaded_trip
Annual_trucking_distance_with_natural_aggregates = 2*Annual_trips_with_natural_aggregates*Average_
  distance_traveled_one_way_with_natural_aggregates
Annual_trucking_distance_with_recycled_material = 2*Average_distance_from_milled_road_to_recycling_
  plant*Annual_trips_with_recycled_material
Average_distance_from_milled_road_to_recycling_plant = 4
Average_distance_traveled_one_way_with_natural_aggregates = 126.2*EXP(-3*Natural_aggregate_stock/
  INIT(Natural_aggregate_stock))
Average_pavement_life = 10 (**years**)
Diesel_use_rate = 1/5
Fraction_Indicated_Aggregate_use_Recycled = 0
Fraction_of_roads_rehabilitated_per_year = 0.7
Indicated_aggregate_use_for_new_roads = Paving_of_new_roads*Aggregate_use_per_km_new_roads
Indicated_aggregate_use_for_rehabilitation = Aggregate_use_per_km_rehabilitated*Paving_for_rehabilitation
Indicated_Aggregate_Use_Recycled = Fraction_Indicated_Aggregate_use_Recycled*Indicated_aggregate_use_
  for_rehabilitation
Indicated_paving_for_rehabilitation = Roads_needing_rehabilitation*Fraction_of_roads_rehabilitated_per_year
Milled_material__production_rate = Milling_Coefficient*Indicated_Aggregate_Use_Recycled
Milling_Coefficient = 0.95
New_Road__Cancellation_Rate = 0
Rehab_cancellation_rate = 0
Required_new_roads_per_year = (Desired_km_roads-Total_km_of_roads)/Time_to_change_km_of_roads
Tons_per_loaded_trip = 25
Total_km_of_roads = Roads_in_top_condition+Roads_needing_rehabilitation

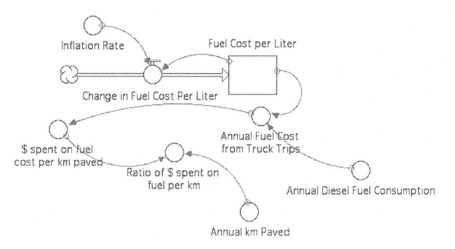

**FIGURE 9.7** Economy sector.

---

## TABLE 9.5
## Equations for Economy Sector

Fuel_Cost_per_Liter(t) = Fuel_Cost_per_Liter(t − dt) + (Change_in_Fuel_Cost_Per_Liter) * dt
INIT Fuel_Cost_per_Liter = 1

*Inflows:*

Change_in_Fuel_Cost_Per_Liter = Inflation_Rate*Fuel_Cost_per_Liter
Inflation_factor(t) = Inflation_factor(t − dt) + (Change_in_inflation_factor) * dt
INIT Inflation_factor = 1

*Inflows:*

Change_in_inflation_factor = Inflation_factor*Inflation_Rate
$_spent_on_fuel_cost_per_km_paved = Annual_Fuel_Cost_from_Truck_Trips/Annual_km_paved
Annual_diesel_fuel_consumption = (Annual_trucking_distance_with_recycled_material+Annual_
    trucking_distance_with_natural_aggregates)*Diesel_use_rate
Annual_Fuel_Cost_from_Truck_Trips = Annual_diesel_fuel_consumption*Fuel_Cost_per_Liter
Annual_km_paved = Paving_for_rehabilitation+Paving_of_new_roads
Inflation_Rate = 0.01
Percentage_Increase__in_Fuel_Cost = 100*(Annual_Fuel_Cost_from_Truck_Trips-INIT(Annual_Fuel_
    Cost_from_Truck_Trips))/init(Annual_Fuel_Cost_from_Truck_Trips)
Ratio_of_$_spent_on__fuel_per_km = ($_spent_on_fuel_cost_per_km_paved) /
    (INIT($_spent_on_fuel_cost_per_km_paved)*Inflation_factor)

---

Next, we consider the environment sector. In this sector, we need to consider the amount of $CO_2$ that is generated by the burning of fuel (which we had determined in the road construction and rehabilitation sector). This requires the consideration of the amount of $CO_2$ generated by burning per unit amount of fuel; however, we should keep in mind that with developments in cleaner technologies, this number will get reduced over time, and hence, it is not a constant. Then, once $CO_2$ is released, it can

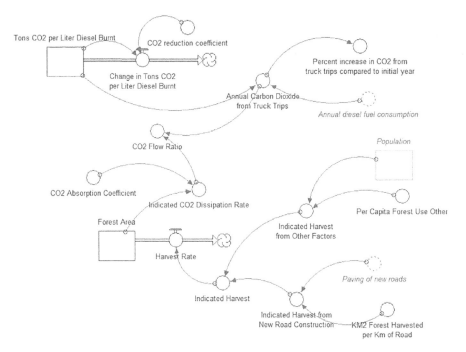

**FIGURE 9.8**   Environment sector.

be absorbed by the forests (trees), whose area will change over time – most likely it will get reduced due to the need for more area for road construction and other human activities. Thus, we need to consider the "harvest" rate of forests for constructing 1 km of road (then we can tie it with the total new roads constructed every year) and per capita per year for non-road-related activities. One way to consider the combined effect of $CO_2$ generation and dissipation will be to use a $CO_2$ flow ratio, which is a ratio of $CO_2$ generated due to burning of fossil fuel to the $CO_2$ absorbed by trees, per year. Thus, we can consider the map and the equations shown in Figure 9.8 and Table 9.6, respectively.

Finally, let us consider the sustainability sector. In this sector, our intent is to determine a Composite Index, that gives us an indication of the sustainability of road construction, with respect to road quality (that affects the quality of life) (engineering consideration), the $CO_2$ flow ratio (environment), and the cost (economy). For road quality, let us consider the actual paving of new roads and rehabilitation and the required (or indicated) paving. The difference between the required and the actual paving will be a shortfall. An increase in the shortfall will lead to a reduction in road quality and a fall in the "score" of the system. An increase in $CO_2$ flow ratio and dollars spent per kilometer of paved roads will also lead to a decrease in the score. Then, if we compare the score at any year with the initial score (initial year), we can evaluate the sustainability of the road construction and rehabilitation work over time. Hence, we propose the map and the equations shown in Figure 9.9 and Table 9.7, respectively.

## TABLE 9.6
### Equations for the Environment Sector

Forest_Area(t) = Forest_Area(t − dt) + (−Harvest_Rate) * dt

INIT Forest_Area = 4,000,000

*Outflows:*

Harvest_Rate = Indicated_Harvest

Population(t) = Population(t − dt)

INIT Population = 1,000,000

kg_$CO_2$_per_Liter_Diesel_Burnt(t) = kg_$CO_2$_per_Liter_Diesel_Burnt(t − dt) + (−Change_in_kg_$CO_2$_per_Liter_Diesel_Burnt) * dt

INIT kg_$CO_2$_per_Liter_Diesel_Burnt = 2.65

*Outflows:*

Change_in_kg_$CO_2$_per_Liter_Diesel_Burnt = $CO_2$_reduction_coefficient*kg_$CO_2$_per_Liter_Diesel_Burnt

Paving_of_new_roads = Required_new_roads_per_year-New_Road__Cancellation_Rate

Annual_Carbon_Dioxide_from_Truck_Trips = kg_$CO_2$_per_Liter_Diesel_Burnt*Annual_diesel_fuel_consumption

Annual_diesel_fuel_consumption = (Annual_trucking_distance_with_recycled_material+Annual_trucking_distance_with_natural_aggregates)*Diesel_use_rate

$CO_2$_Absorption_Coefficient = 439.8

$CO_2$_Flow_Ratio = Annual_Carbon_Dioxide_from_Truck_Trips/Indicated_$CO_2$_Dissipation_Rate

$CO_2$_reduction_coefficient = 0

Indicated_$CO_2$_Dissipation_Rate = IF (Forest_Area > 0) THEN (Forest_Area*$CO_2$_Absorption_Coefficient) ELSE 0.01

Indicated_Harvest = Indicated_Harvest_from_New_Road_Construction+Indicated_Harvest_from_Other_Factors

Indicated_Harvest_from_New_Road_Construction = Paving_of_new_roads*KM2_Forest_Harvested_per_Km_of_Road

Indicated_Harvest_from_Other_Factors = Population*Per_Capita_Forest_Use_Other

KM2_Forest_Harvested_per_Km_of_Road = 0.01

Percent_increase_in_$CO_2$_from_truck_trips_compared_to_initial_year = 100*(Annual_Carbon_Dioxide_from_Truck_Trips-INIT(Annual_Carbon_Dioxide_from_Truck_Trips))/INIT(Annual_Carbon_Dioxide_from_Truck_Trips)

Per_Capita_Forest_Use_Other = 0.03

## COMPLEX PROBLEM – COMBINING THE SECTORS AND RUNNING THE MODEL

This is the fun part. You can put all the sectors together and then run them. But before we do that, we need to consider some important factors.

When we say that we want to run the model, what exactly are we trying to do? Keep in mind, our ultimate objective for modeling is to get some insight into the problem and find some answers, and NOT to get a PERFECT model of the system. Let us review the objective.

We would like to evaluate the sustainability of the road construction and rehabilitation work over time. This means we need to run the model over a time period. Let us consider a time period of 50 years. Next, we need to fix a time interval over which

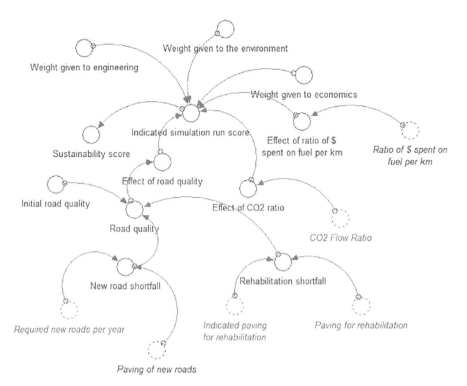

**FIGURE 9.9** Sustainability sector.

the model will integrate (step by step). It is not worth choosing a very small time step – in fact, the time step has nothing to do with actual time, it is required just for numerical integration. So, we should try to select a time step that is just small enough to give us a "smooth" (arguable) plot of the variables (compared with a jagged plot), and not any smaller, to avoid unnecessary computing resources. In this case, let us consider 0.25 year as the time step.

We select a few variables that are important for the model. For example, the fraction of roads rehabilitated, the percentage of aggregates recycled, and the effect of cleaner engine technology are very likely important factors that affect the sustainability of the system, which we evaluate in terms of the sustainability score. We would like to see plots of the sustainability score for different combinations of the above factors. Therefore, we should run the model for the different values of the important variables and evaluate the results. This is the most important benefit of running the model – to understand which factor(s) is (are) more important than the others. We can, therefore, focus on those factors in our work and policy, such that both are effective. We can call these factors (as well as the timing of changes in these factors) "*leverage*" factors, which can have significant effects on the end result. Also, since we are considering the effects of a variety of factors (engineering, environment, and economy) on the end result, we are using an *empathetic* view, where we are considering the view point of the economist, the engineer, and the environmentalist. We are considering the fact that road construction leads to deterioration of the

## TABLE 9.7
## Equations for the Sustainability Sector

Paving_for_rehabilitation = Indicated_paving_for_rehabilitation-Rehab_cancellation_rate

Paving_of_new_roads = Required_new_roads_per_year-New_Road__Cancellation_Rate

$CO_2$_Flow_Ratio = Annual_Carbon_Dioxide_from_Truck_Trips/Indicated_$CO_2$_Dissipation_Rate

Effect_of_$CO_2$_ratio = 3.3 − 1.6*$CO_2$_Flow_Ratio

Effect_of_ratio_of_$_spent_on_fuel_per_km = 2.2 − 0.08*Ratio_of_$_spent_on__fuel_per_km

Effect_of_road_quality = −2.1*(Road_quality^2) + 4.26*Road_quality − 0.16

Indicated_paving_for_rehabilitation = Roads_needing_rehabilitation*Fraction_of_roads_rehabilitated_per_year

Indicated_simulation_run score = ( (Effect_of_road_quality ^ Weight_given_to_engineering ) * (Effect_of_ratio_of_$_spent_on_fuel_per_km ^ Weight_given_to_economics) * ( Effect_of_$CO_2$_ratio^ Weight_given_to_the_environment ) ) ^ (1/(Weight_given_to_economics + Weight_given_to_engineering + Weight_given_to_the_environment ) )

Initial_road_quality = 1

New_road_shortfall = (Required_new_roads_per_year-Paving_of_new_roads)/Required_new_roads_per_year

Ratio_of_$_spent_on__fuel_per_km = ($_spent_on_fuel_cost_per_km_paved)/(INIT($_spent_on_fuel_cost_per_km_paved)*Inflation_factor)

Rehabilitation_shortfall = (Indicated_paving_for_rehabilitation-Paving_for_rehabilitation)/Indicated_paving_for_rehabilitation

Required_new_roads_per_year = (Desired_km_roads-Total_km_of_roads)/Time_to_change_km_of_roads

Road_quality = IF TIME=0 THEN Initial_road_quality ELSE(Initial_road_quality*((1−New_road_shortfall)^0.5)*((1−Rehabilitation_shortfall)^0.5))

Sustainability_score = 100*Indicated_simulation_run_score/INIT(Indicated_simulation_run_score)

Weight_given_to_economics = 0.33

Weight_given_to_engineering = 0.34

Weight_given_to_the_environment = 0.33

environment, but at the same time, we are also acknowledging the fact that without roads and properly rehabilitated roads (good quality roads), the quality of life of citizens falls, and hence, the sustainability score also falls.

The model that we have developed through the different sectors is now ready for running for some trial values. Figure 9.10 shows the final model. Now let us evaluate the effect of recycling of aggregates on the sustainability score.

For recycling percentages, Figure 9.11 shows the sustainability score for 0%, 10%, 40%, and 80% of recycling. Even though the score improves very slightly in the beginning, because of better roads (engineering criterion), the sustainability score deteriorates over time; however, the scores for the cases with recycling remain higher than that of no recycling, and the higher the recycling percentage, the higher the score at any point of time. Hence, the model is able to capture the beneficial effect of recycling on the overall sustainability of the system. One can experiment with other variables to evaluate their effects, such as the recycling percentage on the natural aggregate stock (see Figure 9.12).

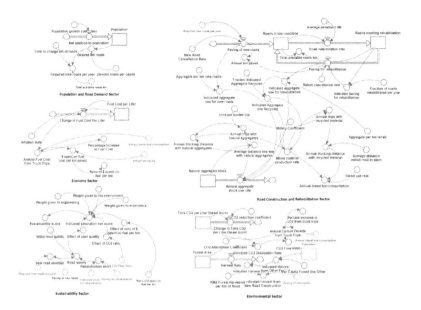

**FIGURE 9.10** Complete model.

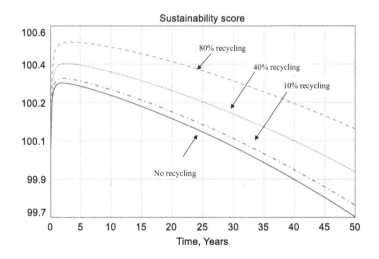

**FIGURE 9.11** Effect of recycling on sustainability score.

# EXAMPLE: A MODEL TO DETERMINE THE EFFECT OF CLIMATE CHANGE ON PAVEMENT LIFE

Climate change is an important topic in all disciplines. Changes in climate, in the form of heavy rainfall, high temperatures, and increase in the number of floods causing hurricanes, can affect pavement performance and conditions in a very detrimental

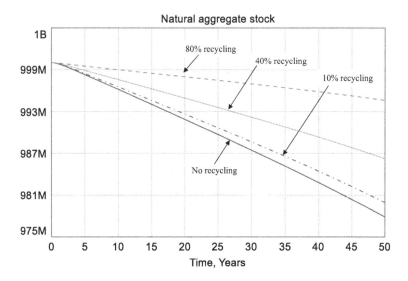

**FIGURE 9.12**   Effect of recycling on the natural aggregate stock.

way (Mallick et al., 2014, 2016). While the effects of maintenance and costs may not be significant in the short term, the effects could be very significant in the long term. To determine the effects of changes in climate on pavement life, a system dynamics model was created with available pavement performance and climate change data, and the effects on various key factors such as pavement life and maintenance cost were evaluated through simulation. A flow chart explains the modeling approach in Figure 9.13. The causal structure is shown in Figure 9.14. The key variables are shown in Table 9.8.

There are four main climate impacts that have been considered: change in maximum air temperature, sea level rise, change in the number of hurricanes (≥category 3) per year, and change in the average annual precipitation. In this example, we analyze a coastal roadway, and the first three of the above factors are considered to cause inundations of the roadway, and the last one is assumed to cause a higher saturation level of the subgrade soil. Accordingly, the modulus values of the surface and subgrade soils are adjusted for those specific conditions. The adjusted modulus values are then used in predicting the life of the pavement (in terms of rutting or fatigue crackling, whichever is critical or shorter), by using regression equations that have been developed for the roadway with Mechanistic Empirical Pavement Design Guide/Software. Finally, the predicted life of the pavement is used to determine the deterioration rate of the pavement over time.

The model was simulated with different maximum air temperature and precipitation increases (sensitivity analyses), and the results are shown in Figures 9.15 and 9.16. It can be clearly seen that as both of those parameters increase, the average pavement life decreases, and the reduction in life is significant. This translates to a higher cost of maintenance since roads will deteriorate faster over time due to climate change impacts.

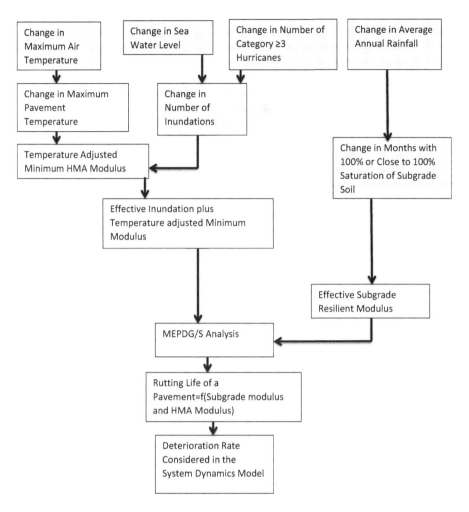

**FIGURE 9.13**  Modeling approach for the effect of climate. (MEPDG/S, Mechanistic Empirical Pavement Design Guide/Software.)

# EXAMPLE: A MODEL TO DETERMINE THE LONG-TERM BENEFITS OF USING WARM MIX ASPHALT, DRY AGGREGATES, AND RECLAIMED ASPHALT PAVEMENT

The production of HMA during pavement maintenance and rehabilitation results in significant use of energy and release of $CO_2$. The use of sustainable technologies, namely, warm mix asphalt (WMA), which uses lower than conventional production temperature, reclaimed asphalt pavements for recycling, and using drier aggregates (which require lower energy to dry them before mixing with asphalt) can result in savings, in terms of energy and emissions (NAPA, 2022). The problem is to determine the long-term benefits of using these technologies.

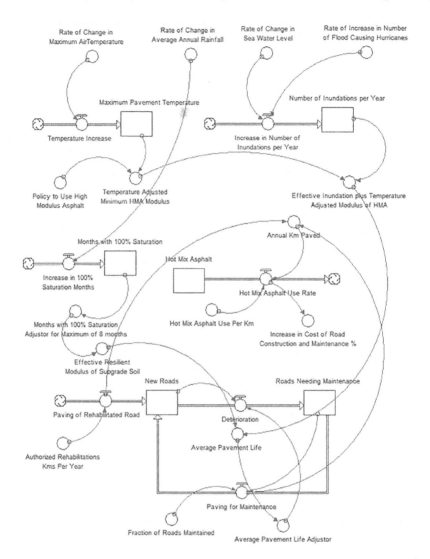

**FIGURE 9.14**   Causal structure of effect of climate change.

Energy consumption and emission of $CO_2$, CO, $SO_2$, and $NO_x$ (all of which can be quantified in terms of equivalent $CO_2$) are the main concerns regarding the impact of the road-building industry on the environment. Some of the practical steps that could be taken to reduce the amount of energy and cut down the amount of emissions are as follows:

1. Reduce the moisture in the aggregate stockpile, since an increase in the moisture content of the aggregates leads to the need for an increased amount of energy to dry the aggregates, which is required to avoid insufficient asphalt binder coating and stripping damage in the field. This can be achieved by keeping the aggregate stockpiles under a cover and/or in a dry place.

## TABLE 9.8
### Key Variables

Hot_Mix_Asphalt(t) = Hot_Mix_Asphalt(t − dt) + (−Hot_Mix_Asphalt_Use_Rate) * dt
INIT Hot_Mix_Asphalt = 1e + 20
*Outflows*:
Hot_Mix_Asphalt_Use_Rate = Annual_Km_Paved*Hot_Mix_Asphalt_Use_Per_Km
Number_of_Inundations_per_Year(t) = Number_of_Inundations_per_Year(t − dt) + (Increase_in_
    Number_of_Inundations_per_Year) * dt
INIT Number_of_Inundations_per_Year = 0.267 + 1 + 0.150
*Inflows*:
Increase_in_Number_of_Inundations_per_Year = 10.695*Rate_of_Change_in_Sea_Water_
    Level+1*Rate_of_Increase_in_Number_of_Flood_Causing_Hurricanes
Roads_Needing_Maintenance(t) = Roads_Needing_Maintenance(t − dt) + (Deterioration − Paving_for_
    Maintenance) * dt
INIT Roads_Needing_Maintenance = 50
*Inflows*:
Deterioration = New_Roads/Average_Pavement_Life_Adjustor
*Outflows*:
Paving_for_Maintenance = Roads_Needing_Maintenance*Fraction_of_Roads_Maintained
Maximum_Pavement_Temperature(t) = Maximum_Pavement_Temperature(t − dt) + (Temperature_
    Increase) * dt
INIT Maximum_Pavement_Temperature = 58
*Inflows*:
Temperature_Increase = 0.78*Rate_of_Change_in_Maximum_AirTemperature
Months_with_100%_Saturation(t) = Months_with_100%_Saturation(t − dt) + (Increase_in_100%_
    Saturation_Months) * dt
INIT Months_with_100%_Saturation = 2
*Inflows*:
Increase_in_100%_Saturation_Months = (0.0071*Rate_of_Change_in_Average_Annual_Rainfall)
New_Roads(t) = New_Roads(t − dt) + (Paving_of_Rehabilitated_Road + Paving_for_
    Maintenance − Deterioration) * dt
INIT New_Roads = 0
*Inflows*:
Paving_of_Rehabilitated_Road = Authorized_Rehabilitations_Kms_Per_Year
Paving_for_Maintenance = Roads_Needing_Maintenance*Fraction_of_Roads_Maintained
*Outflows*:
Deterioration = New_Roads/Average_Pavement_Life_Adjustor
Annual_Km_Paved = Paving_of_Rehabilitated_Road+Paving_for_Maintenance
Authorized_Rehabilitations_Kms_Per_Year = 2
Average_Pavement_Life = -25.574+0.000551*Effective_Resilient_Modulus_of_Subgrade_
    Soil+0.0000452*Effective_Inundation_plus_Temperature_Adjusted_Modulus_of_HMA
Average_Pavement_Life_Adjustor = If (Average_Pavement_Life > 1) THEN (Average_Pavement_Life)
    ELSE (1)
Effective_Inundation_plus_Temperature_Adjusted_Modulus_of_HMA = Temperature_Adjusted_
    Minimum_HMA_Modulus-0.03*Number_of_Inundations_per_Year*INIT(Temperature_Adjusted_
    Minimum_HMA_Modulus)
Effective_Resilient_Modulus_of_Subgrade_Soil = 26,500*((12-Months_with_100%_Saturation_
    Adjustor_for_Maximum_of_8_months)/12)+8,690*(Months_with_100%_Saturation_Adjustor_for_
    Maximum_of_8_months/12)

(*Continued*)

## TABLE 9.8 (*Continued*)
## Key Variables

Fraction_of_Roads_Maintained = 0.1

Hot_Mix_Asphalt_Use_Per_Km = 1,272

Increase_in_Cost_of_Road_Construction_and_Maintenance_% = 100*(Hot_Mix_Asphalt_Use_Rate-INIT(Hot_Mix_Asphalt_Use_Rate))/INIT(Hot_Mix_Asphalt_Use_Rate)

Months_with_100%_Saturation_Adjustor_for_Maximum_of_8_months = IF(Months_with_100%_Saturation<=8)THEN(Months_with_100%_Saturation)ELSE(8)

Policy_to_Use_High_Modulus_Asphalt = 0

Rate_of_Change_in_Average_Annual_Rainfall = 2.8189

Rate_of_Change_in_Maximum_AirTemperature = 0.024

Rate_of_Change_in_Sea_Water_Level = 0.0093

Rate_of_Increase_in_Number_of_Flood_Causing_Hurricanes = 0.000556

Temperature_Adjusted_Minimum_HMA_Modulus = 1588003-16012*Maximum_Pavement_Temperature+4000*Maximum_Pavement_Temperature*Policy_to_Use_High_Modulus_Asphalt

Fraction_of_Roads_Maintained = 0.1

Hot_Mix_Asphalt_Use_Per_Km = 1,272

Increase_in_Cost_of_Road_Construction_and_Maintenance_% = 100*(Hot_Mix_Asphalt_Use_Rate-INIT(Hot_Mix_Asphalt_Use_Rate))/INIT(Hot_Mix_Asphalt_Use_Rate)

Months_with_100%_Saturation_Adjustor_for_Maximum_of_8_months = IF(Months_with_100%_Saturation<=8)THEN(Months_with_100%_Saturation)ELSE(8)

Policy_to_Use_High_Modulus_Asphalt = 0

Rate_of_Change_in_Average_Annual_Rainfall = 2.8189

Rate_of_Change_in_Maximum_AirTemperature = 0.024

Rate_of_Change_in_Sea_Water_Level = 0.0093

Rate_of_Increase_in_Number_of_Flood_Causing_Hurricanes = 0.000556

Temperature_Adjusted_Minimum_HMA_Modulus = 1588003-16012*Maximum_Pavement_Temperature+4000*Maximum_Pavement_Temperature*Policy_to_Use_High_Modulus_Asphalt

2. Reduce heat loss during production, as the amount of heat loss in the dryer drum, during the drying of the aggregates can lead to wastage of fuel; better insulation that leads to a reduction in wastage of heat can result in significant savings in energy and reduction in emissions.

3. Reduce the temperature of asphalt mixes, as the amount of emissions is directly proportional to the amount of energy that is used for the production of the mix. Such reduction can be accomplished by using WMA.

4. Use gas fuel instead of liquid fuel, since the use of gas fuels generally results in the reduction of emissions.

5. Reduce the amount of mix per kilometer of pavement, since the amount of energy used and emissions are proportional to the amount of materials used.

Using these five parameters as variables, a system dynamics model was created and simulated for different values. A schematic of the modeling approach is shown in Figure 9.17. The causal structure is shown in Figure 9.18. The key variables are shown in Table 9.9.

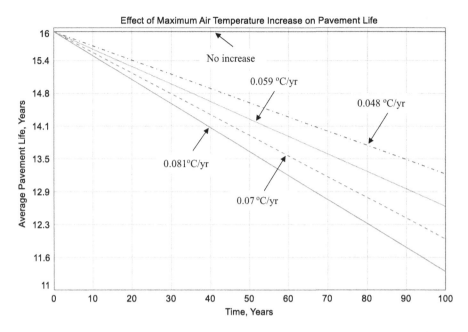

**FIGURE 9.15**   Effect of maximum air temperature increase on the average pavement life.

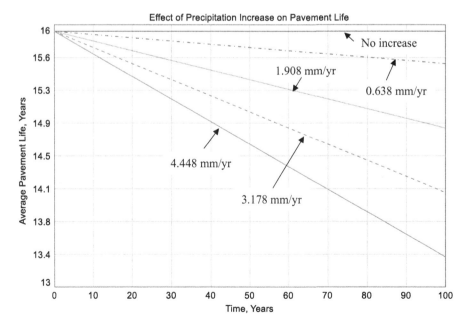

**FIGURE 9.16**   Effect of precipitation increase on the average pavement life.

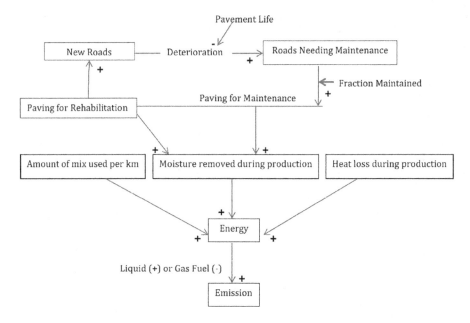

**FIGURE 9.17**    Schematic of modeling approach.

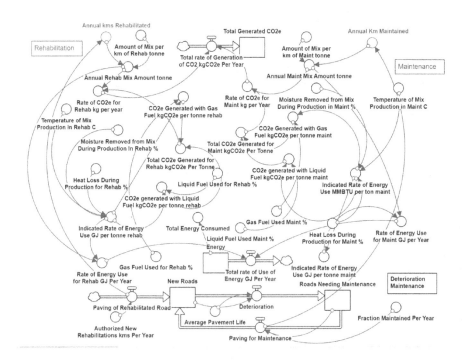

**FIGURE 9.18**    Causal structure.

## TABLE 9.9
## Key Variables

New_Roads(t) = New_Roads(t − dt) + (Paving_of_Rehabilitated_Road + Paving_for_
   Maintenance − Deterioration) * dt
INIT New_Roads = 0
*Inflows*:
Paving_of_Rehabilitated_Road = Authorized_New_Rehabilitations_kms_Per_Year
Paving_for_Maintenance = Roads_Needing_Maintenance*Fraction_Maintained_Per_Year
*Outflows*:
Deterioration = New_Roads/Average_Pavement_Life
Roads_Needing_Maintenance(t) = Roads_Needing_Maintenance(t - dt) + (Deterioration - Paving_for_
   Maintenance) * dt
INIT Roads_Needing_Maintenance = 1,000
*Inflows*:
Deterioration = New_Roads/Average_Pavement_Life
*Outflows*:
Paving_for_Maintenance = Roads_Needing_Maintenance*Fraction_Maintained_Per_Year
Energy(t) = Energy(t − dt) + (−Total_rate_of_Use_of_Energy_GJ_Per_Year) * dt
INIT Energy = 10,000,000,000
*Outflows*:
Total_rate_of_Use_of_Energy_GJ_Per_Year = Rate_of_Energy_Use_for_Rehab_GJ_Per_Year+Rate_
   of_Energy_Use_for_Maint_GJ_per_Year
Total_Generated_CO$_2$e(t) = Total_Generated_CO$_2$e(t − dt) + (Total_rate_of_Generation_of_CO$_2$_
   kgCO$_2$e_Per_Year) * dt
INIT Total_Generated_CO$_2$e = 0
*Inflows*:
Total_rate_of_Generation_of_CO$_2$_kgCO$_2$e_Per_Year = Rate_of_CO$_2$e_for_Maint_kg_per_Year+Rate_
   of_CO$_2$e_for_Rehab_kg_per_year
Amount_of_Mix_per_km_of_Maint_tonne = 15,000
Amount_of_Mix_per_km_of_Rehab_tonne = 20,000
Annual_kms_Rehabilitated = Paving_of_Rehabilitated_Road
Annual_Km_Maintained = Paving_for_Maintenance
Annual_Maint_Mix_Amount_tonne = Amount_of_Mix_per_km_of_Maint_tonne*Annual_Km_
   Maintained
Annual_Rehab_Mix_Amount_tonne = Amount_of_Mix_per_km_of_Rehab_tonne*Annual_kms_
   Rehabilitated
Authorized_New_Rehabilitations_kms_Per_Year = 5000
Average_Pavement_Life = 11
CO2e_Generated_with_Gas_Fuel_kgCO2e_per_tonne_maint = (Indicated_Rate_of_Energy_Use_
   MMBTU_per_ton_maint*55.37-0.630)*(Gas_Fuel_Used_Maint_%)/100
CO2e_Generated_with_Gas_Fuel_kgCO2e_per_tonne_rehab = (Indicated_Rate_of_Energy_Use_GJ_
   per_tonne_rehab*55.37-0.630)*(Gas_Fuel_Used_for_Rehab_%)/100
CO2e_generated_with_Liquid_Fuel_kgCO2e_per_tonne_maint = (Indicated_Rate_of_Energy_Use_
   MMBTU_per_ton_maint*75.895-0.800)*(Liquid_Fuel_Used_Maint_%)/100
CO2e_generated_with_Liquid_Fuel_kgCO2e_per_tonne_rehab = (Indicated_Rate_of_Energy_Use_GJ_
   per_tonne_rehab*75.895-0.800)*(Liquid_Fuel_Used_for_Rehab_%)/100
Fraction_Maintained_Per_Year = 0.3
Gas_Fuel_Used_for_Rehab_% = 50

*(Continued)*

## TABLE 9.9 (*Continued*)
## Key Variables

Gas_Fuel_Used_Maint_% = 50

Heat_Loss_During_Production_for_Maint_% = 3

Heat_Loss_During_Production_for_Rehab_% = 3

Indicated_Rate_of_Energy_Use_GJ_per_tonne_maint = Indicated_Rate_of_Energy_Use_MMBTU_per_ton_maint*1.05587*1.1

Indicated_Rate_of_Energy_Use_GJ_per_tonne_rehab = (-0.005+0.001*(Temperature_of_Mix_Production_In_Rehab_C)+0.026*(Moisture_Removed_from_Mix_During_Production_In_Rehab_%))*(1+0.01*(Heat_Loss_During_Production_for_Rehab_%-12))

Indicated_Rate_of_Energy_Use_MMBTU_per_ton_maint = (-0.005+0.001*(Temperature_of_Mix_Production_in_Maint_C)+0.026*(Moisture_Removed_from_Mix_During_Production_in_Maint_%))*(1+0.01*(Heat_Loss_During_Production_for_Maint_%-12))

Liquid_Fuel_Used_for_Rehab_% = 50

Liquid_Fuel_Used_Maint_% = 50

Moisture_Removed_from_Mix_During_Production_in_Maint_% = 4

Moisture_Removed_from_Mix_During_Production_In_Rehab_% = 4


Rate_of_CO2e_for_Maint_kg_per_Year = Total_CO2e_Generated_for_Maint_kgCO2e_Per_Tonne*Annual_Maint_Mix_Amount_tonne

Rate_of_CO2e_for_Rehab_kg_per_year = Total_CO2e_Generated_for_Rehab_kgCO2e_Per_Tonne*Annual_Rehab_Mix_Amount_tonne

Rate_of_Energy_Use_for_Maint_GJ_per_Year = Indicated_Rate_of_Energy_Use_GJ_per_tonne_maint*Annual_Maint_Mix_Amount_tonne

Rate_of_Energy_Use_for_Rehab_GJ_Per_Year = Indicated_Rate_of_Energy_Use_GJ_per_tonne_rehab*Annual_Rehab_Mix_Amount_tonne

Temperature_of_Mix_Production_in_Maint_C = 150

Temperature_of_Mix_Production_In_Rehab_C = 150

Total_CO2e_Generated_for_Maint_kgCO2e_Per_Tonne = CO2e_Generated_with_Gas_Fuel_kgCO2e_per_tonne_maint+CO2e_generated_with_Liquid_Fuel_kgCO2e_per_tonne_maint

Total_CO2e_Generated_for_Rehab_kgCO2e_Per_Tonne = CO2e_Generated_with_Gas_Fuel_kgCO2e_per_tonne_rehab+CO2e_generated_with_Liquid_Fuel_kgCO2e_per_tonne_rehab

Total_Energy_Consumed = INIT(Energy)-Energy

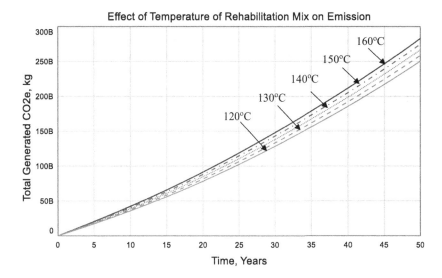

**FIGURE 9.19**   Effect of temperature of mix used during rehabilitation on emissions.

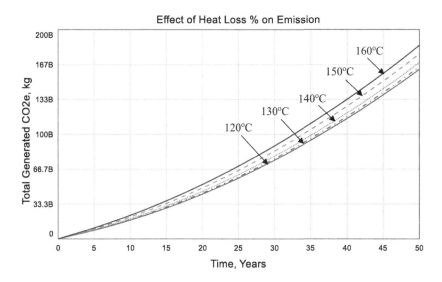

**FIGURE 9.20**   Effect of heat loss from dryer, %, during the production of mix used during rehabilitation on emissions.

Now we can simulate the model to explore the impact of the change of different variables. For example, the results of a sensitivity analysis for the temperature of the mix used in rehabilitation are shown in Figure 9.19. It shows that as the temperature decreases, the emissions decrease. Similarly, as the heat loss during production of the mix during rehabilitation decreases, the emissions also decrease (see Figure 9.20).

## REFERENCES

Mallick, R. B., Radzicki, M. J., Daniel, J. S., Jacobs, J. M. (2014). Use of system dynamics to understand the long term impact of climate change on pavement performance and maintenance cost. Transportation research Record: Journal of the Transportation Research Board, 2455, Washington DC.

Mallick, R. B., Jacobs, J. M., Miller, B. J., Daniel, J. S., and Kirshen, P. (2015). Understanding the impact of climate change on pavements with CMIP5, system dynamics and simulation. *International Journal of Pavement Engineering*. 19(8), 697–705. DOI: 10.1080/10298436.2016.1199880

National Asphalt Pavement Association (NAPA) 2022. Asphalt Pavement Industry Goals for Climate Stewardship: Toward Net Zero Carbon Emissions https://www.asphaltpavements.org/climate/industry-goals

# 10 Probabilistic Simulation with System Dynamics

This chapter presents and demonstrates the use of statistical analysis, namely, Monte Carlo simulation along with system dynamics to derive realistic numerical estimates of critical parameters. Note that while the other models focused primarily on trends, here, the methodology shows explicitly a technique to determine numerical results.

Climate change–related increase in maximum air temperature and annual rainfall has major negative impacts on pavement performance and reduces their lives, as we saw in the last chapter. As a consequence, the mileage of roads that would require rehabilitation at any point of time in the future would tend to increase. To make roads resilient to climate change impacts, agencies can use better materials/more durable materials such as high-modulus asphalt. However, any such use involves additional investment, which would require justification, specifically when the uncertainties in climate change–related impacts are considered.

In this chapter, we will explore a framework (Mallick and Nazarian, 2018) that builds on a system dynamics model to predict the mileage of roads that require rehabilitation due to climate change and then use Monte Carlo simulation to evaluate the impact of the statistical uncertainties, to derive at a probabilistic estimate of justifiable investment. The main principle of this approach is that the level of investment is justified that will result in a return (by reducing the cost of additional rehabilitation due to climate change impact) which is the same as that can be obtained from a regular monetary investment at a reasonable interest rate.

In this example, the network is assumed to be of 160 lane-km (100 lane-miles) of flexible pavement. Each lane-mile has a 3.6 m (12 ft) wide main lane and a 1.8 m (6 ft) wide shoulder. The analysis period is considered to be 15 years, which is a typical pavement design period and also sufficiently long that is needed for climate change signals to be "detected" by climate data (Hawkins and Sutton, 2009). The rehabilitation work in 15 years consists of a 75 mm (3-in) thick layer of hot mix asphalt, costing $104,160 per lane-km ($168,000 per lane-mile).

The analysis is conducted through the following steps.

1. Consider rates of change in maximum air temperature and precipitation (Mallick et al., 2016); see Table 10.1.
2. Develop the system dynamics model; see Figure 10.1.
3. Run the system dynamics model with different rates of changes in maximum air temperature and precipitation, and develop a regression model; see Table 10.2.
4. Determine the range of extra rehabilitation costs for the climate change–related impacts; see Figure 10.2 for a time span of 15 years.

## TABLE 10.1
## Rate of Change in Maximum Air Temperature and Precipitation (Mallick et al., 2016)

| Rate of Change, Per Year | Mean | Standard Deviation |
|---|---|---|
| Maximum air temperature, °C per year | 0.063 | 0.013 |
| Rainfall, mm per year | 1.406 | 0.45 |

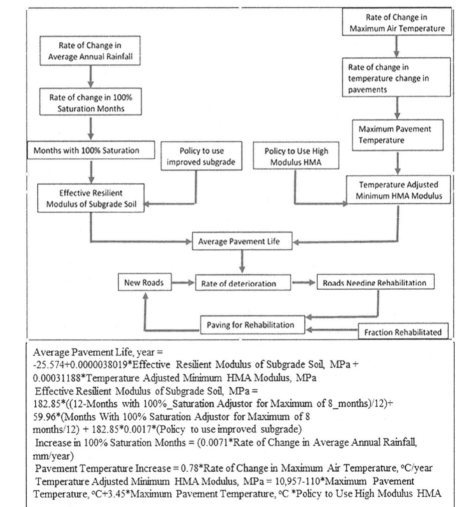

Average Pavement Life, year =
-25.574+0.0000038019*Effective Resilient Modulus of Subgrade Soil, MPa +
0.00031188*Temperature Adjusted Minimum HMA Modulus, MPa
Effective Resilient Modulus of Subgrade Soil, MPa =
182.85*((12-Months with 100%_Saturation Adjustor for Maximum of 8_months)/12)+
59.96*(Months With 100% Saturation Adjustor for Maximum of 8
months/12) + 182.85*0.0017*(Policy to use improved subgrade)
Increase in 100% Saturation Months = (0.0071*Rate of Change in Average Annual Rainfall,
mm/year)
Pavement Temperature Increase = 0.78*Rate of Change in Maximum Air Temperature, °C/year
Temperature Adjusted Minimum HMA Modulus, MPa = 10,957-110*Maximum Pavement
Temperature, °C+3.45*Maximum Pavement Temperature, °C *Policy to Use High Modulus HMA

**FIGURE 10.1**   System dynamics model showing the key parameters.

## TABLE 10.2
## Regression Equations Developed in the Current Study for Percentage of Roads Requiring Rehabilitation at Different Years from the Simulation of the System Dynamic Model

| Year | Intercept | Coefficient for X1, C1 | Coefficient for X2, C2 |
|------|-----------|------------------------|------------------------|
| 0 | 0 | 0 | 0 |
| 5 | 9.90 | 1.04 | 0.01 |
| 10 | 10.95 | 6.49 | 0.06 |
| 15 | 11.72 | 14.52 | 0.14 |
| 20 | 12.28 | 24.43 | 0.25 |
| 25 | 12.68 | 35.74 | 0.36 |
| 30 | 12.97 | 48.12 | 0.49 |
| 35 | 13.17 | 61.28 | 0.63 |
| 40 | 13.31 | 75.04 | 0.77 |
| 45 | 13.40 | 89.24 | 0.91 |
| 50 | 13.45 | 103.76 | 1.06 |

Percentage of roads requiring rehabilitation = Intercept + C1*X1 + C2*X2.
X1, rate of change in max. air temperature; X2, rate of change in annual precipitation.

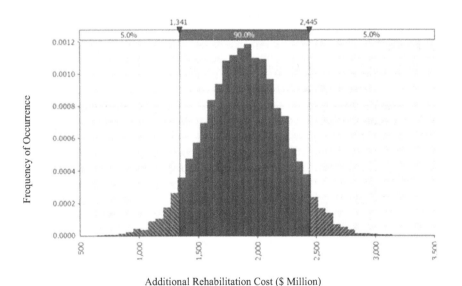

Additional Rehabilitation Cost ($ Million)

**FIGURE 10.2** Frequency distribution of climate change–related extra cost for maintaining roads in 15 years.

5. Compute the extra percentage of roads requiring rehabilitation; see Table 10.3.
6. Run the system dynamics model with high-modulus asphalt properties and determine the percent of roads requiring rehabilitation; compare this with roads requiring rehabilitation for conventional asphalt; see Table 10.4.
7. From step 6, compute the cost reduction (benefit) in rehabilitation by using high-modulus asphalt; see Figure 10.3 (Button et al., 1987, Carpenter and VanDam, 1987, Russo et al., 2023).
8. Show the percent of roads requiring rehabilitation for the different conditions in 15 years; see Figure 10.4.
9. Now conduct a Monte Carlo simulation as an example to show the net return of investing $500 in 15 years; see Figure 10.5.
10. Now compare the cost reduction benefits of using high-modulus asphalt (both deterministic and probabilistic) with the net return of investing different amounts of money at a reasonable interest rate; see Figure 10.6.
11. Compare the benefits (by using the high-modulus asphalt) with the net returns (investment) for 5th to 95th percentile; see Figure 10.7.

So, what did we learn from constructing and simulating this model? First, the process of modeling helps us understand the linkage between the different variables in the problem and the impact of uncertainty. Second, the process results in a rational framework, which can be used by agencies to decide on investments for keeping roads in good condition in the face of climate change impacts. Next, the model provides a method to obtain confidence intervals regarding benefits for specific amounts of investments, with a consideration of uncertainties in climate change predictions.

---

**TABLE 10.3**
**Extra Percentage of Roads Requiring Rehabilitation**

| | Percent of Roads Needing Rehabilitation | | |
|---|---|---|---|
| Time | Not Considering Climate Change | Considering Climate Change | Extra Percentage of Roads Requiring Rehabilitation |
| 0 | 0 | 0 | 0 |
| 5 | 9.90 | 9.90 + 1.04(X1) + 0.010(X2) | 1.04*X1 + 0.01*X2 |
| 10 | 10.96 | 10.95 + 6.49(X1) + 0.06(X2) | 6.49*X1 + 0.07*X2 |
| 15 | 11.74 | 11.72 + 14.52(X1) + 0.14(X2) | 14.52*X1 + 0.15*X2 |
| 20 | 12.32 | 12.28 + 24.43(X1) + 0.25(X2) | 24.43*X1 + 0.25*X2 |
| 25 | 12.76 | 12.68 + 35.74(X1) + 0.36(X2) | 35.75*X1 + 0.37*X2 |
| 30 | 13.08 | 12.97 + 48.12(X1) + 0.49(X2) | 48.12*X1 + 0.49*X2 |
| 35 | 13.32 | 13.17 + 61.28(X1) + 0.63(X2) | 61.29*X1 + 0.63*X2 |
| 40 | 13.50 | 13.31 + 75.04(X1) + 0.77(X2) | 75.05*X1 + 0.77*X2 |
| 45 | 13.64 | 13.40 + 89.24(X1) + 0.91(X2) | 89.24*X1 + 0.92*X2 |
| 50 | 13.74 | 13.45 + 103.76(X1) + 1.06(X2) | 103.77*X1 + 1.07*X2 |

X1, rate of change in maximum air temperature, C/yr; X2, rate of change in annual rainfall, mm/yr.

## TABLE 10.4
### Regression Equations from Results of Simulation of System Dynamics Model

#### High-Modulus Asphalt

| Time, at Year | X1 | X2 | Percent of Roads Requiring Rehabilitation, Y |
|---|---|---|---|
| 15 | 0.0238 | 1.406 | 9.89 |
| | 0.063 | 1.406 | 10.04 |
| | 0.102 | 1.406 | 10.36 |
| | 0.063 | 0.056 | 9.98 |
| | 0.063 | 1.406 | 10.04 |
| | 0.063 | 2.756 | 10.13 |

Regression equation: $Y = 9.61 + 6.03X1 + 0.06X2$.
X1, rate of change in maximum air temperature, C per year; X2, rate of change in Annual rainfall, mm per year.

#### Conventional Pavement

| Time, at Year | X1 | X2 | Percent of Roads Requiring Rehabilitation, Y |
|---|---|---|---|
| 15 | 0.0238 | 1.406 | 12.28 |
| | 0.063 | 1.406 | 12.84 |
| | 0.102 | 1.406 | 13.42 |
| | 0.063 | 0.056 | 12.64 |
| | 0.063 | 1.406 | 12.84 |
| | 0.063 | 2.756 | 13.05 |

Regression equation: $Y = 11.72 + 14.52X1 + 0.15X2$.
X1, rate of change in maximum air temperature, C per year; X2, rate of change in annual rainfall, mm per year.

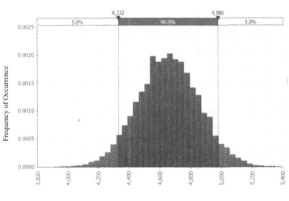

Cost Reduction ($ Million)

FIGURE 10.3 Frequency distribution of benefits of using high-modulus asphalt in 15 years; the benefits are realized by not having to spend to maintain roads affected by climate change.

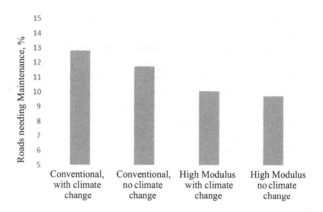

**FIGURE 10.4**    Roads needing maintenance for different types of materials and scenarios in 15 years.

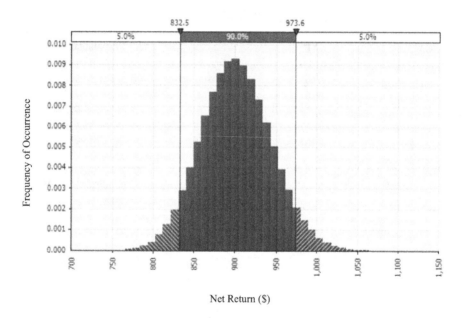

Net Return ($)

**FIGURE 10.5**    Example of Monte Carlo simulation of investment of $500 in 15 years.

The framework also provides the users the scope of using their relevant data, from experiments or experience. Finally, the simulation gives us valuable insights into the long-term behavior of the system, such as the significant effect of climate change on pavement life, the beneficial effects of using high-modulus materials, and the significant effect of the variability/uncertainty on the expected benefits from additional investments.

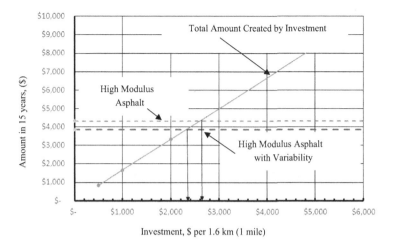

**FIGURE 10.6** Determination of appropriate amount of investment; the plots show fifth percentile data; the slanted straight line represents the accumulated amount due to different amounts of interest; the horizontal lines indicate the benefit of using high-modulus asphalt in 15 years; wherever the two plots intersect, the point of intersection of the slanted horizontal lines indicates the appropriate amount that should be invested to achieve the same benefit of using a high-modulus asphalt as one would expect from a regular interest paying investment in 15 years. Here, assuming no variability, the appropriate investment (that is the amount above what is generally spent on conventional pavements) amount is approximately $2,600 per 1.6 km (1 mile) of the road.

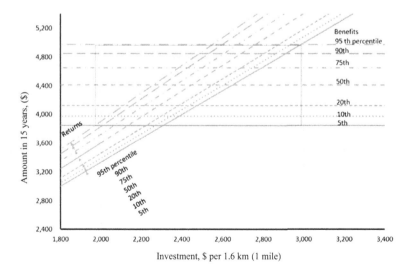

**FIGURE 10.7** Range of benefits and returns for different investments for different percentiles for the case of the high-modulus asphalt with variability; use these data to determine the appropriate level of investment for constructing resilient pavements, for any specific percentiles for benefit and returns.

# REFERENCES

Button , J. W., Little, D. N., Kirn, Y., and Ahmed, J. (1987). Mechanistic evaluation of selected asphalt additives. *Proceedings of Association of Asphalt Paving Technologists*, 56, 62–90.

Carpenter, S. H. and VanDam, T. (1987). Laboratory performance comparisons of polymer modified and unmodified asphalt concrete mixtures. Transportation Research Record: Journal of the TRB 115, Transportation research Board, Washington, DC.

Hawkins, E. and Sutton, R. (2009). The potential to narrow uncertainty in regional climate predictions. *Bulletin of the American Meteorological Society*, 90, 1095–1107.

Mallick, R. B., Jacobs, J. M., Miller, B. J., Daniel, J. S., and Kirshen, P. (2016). Understanding the impact of climate change on pavements with CMIP5, system dynamics and simulation, *International Journal of Pavement Engineering*, doi: 10.1080/10298436.2016.1199880

Mallick, R. B. and Nazarian, S. (2018). A rational method to determine investment amount for making roadways resilient to a changing climate. *Journal of Infrastructure Systems*, 24(1), 04017049. doi: 10.1061/(ASCE)IS.1943-555X.0000411

Russo, F., Veropalumbo, R., and Oreto, C. (2023). Climate change mitigation investigating asphalt pavement solutions made up of plastomeric compounds. *Resource, Conservation and Recycling*, 189.

# 11 System of Systems

Climate change and extreme weather-related events such as coastal storms/hurricanes pose a major threat to the transportation infrastructure, including coastal airports. Evaluation of vulnerabilities for improvement of resilience has become one of the top priorities of departments of transportation and governments.In this chapter, we will explore resilience of coastal airports against extreme weather or climate change–related events such as coastal storms, hurricanes, and storm surges caused by such hurricanes (Mallick et al., 2018). Any major airport consists of multiple assets such as control tower, runways and electrical substations; each of these can be considered to be a system; and hence, we are going to explore, in a sense, the resilience of a system of systems.

We will explore the cascading effect of failure of one system on the other and hence on the entire system of systems. This will bring out the downtime of each system and the entire system due to coastal flooding or storm surge. What is the usefulness of this modeling? First, we will be able to determine which assets have downtimes or vulnerabilities that are above the acceptable limits; second, we can use the model to "experiment" with different types of inputs to reduce their downtimes and vulnerabilities to acceptable levels. For example, if the downtime of 7 hours for a storm surge of 20 ft is found to be unacceptable, then the adaptive capacity of a specific asset could be raised by installing a flood wall to a level that would reduce its downtime and the overall downtime of the infrastructure to an acceptable level.

The modeling is carried out through the following steps.

1. The critical assets of a major airport need to be identified, for constructing the model:
   a. Runways
   b. ATC tower
   c. Electrical substations
2. Establish the mathematical equations from the definitions of asset integrity (AI), vulnerability, risk, and resilience from the literature.
   First, for AI, refer to Figure 11.1, for a storm surge-related flood level in the airport.
   Single asset (independent)

$$AI_i = 1 - \frac{D_{FL_i}(t)}{D_{FL_i}|FAIL} \qquad (11.1)$$

   Asset i dependence on Asset j

$$Ai_{i,j} = \left(1 - \frac{D_{FL\,i}(t)}{D_{FL\,i}|FAIL}\right)\left(1 - \frac{D_{FL\,j}(t)}{D_{FL\,j}|FAIL}\right) \qquad (11.2)$$

DOI: 10.1201/9781003345596-11

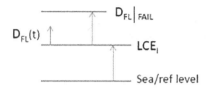

**FIGURE 11.1**   Relationship of the asset-specific lowest critical elevation (LCE) (which is the threshold water level above which the asset is assumed to suffer reduced integrity) and $D_{FL}|FAIL$ measures as used in the simulation model. DFL(t) is the time-dependent water depth above the LCE for a given asset. At $D_{FL}$ operational integrity was reduced to zero.

**TABLE 11.1**

**System Resilience Components (Vugrin et al., 2010)**

| Component | Absorptive Capacity | Adaptive Capacity | Restorative Capacity |
|---|---|---|---|
| Associated with | System impact | System impact | Total recovery effort |
| Distinguishing characteristics of capacity | Considers aspects that automatically manifest after disruption | Considers internal aspects that manifest over time after disruption | Considers ability to affect and repair internal system features |
| Effort required | Automatic/little effort | Internal effort required | External effort often required |
| Measurement of component | Internal measurement | Internal measurement | Exogenous measurement |

Asset i dependence on Assets j and k

$$Ai_{i,j,k} = \left(1 - \frac{D_{FL\,i}(t)}{D_{FL\,i}|FAIL}\right)\left(1 - \frac{D_{FL\,j}(t)}{D_{FL\,j}|FAIL}\right)\left(1 - \frac{D_{FL\,k}(t)}{D_{FL\,k}|FAIL}\right) \quad (11.3)$$

Asset integrity $= 0$, when DFL(t) $\geq$ DFL|FAIL), and increases as the water level falls below DFL|FAIL

For the definitions of resilience, see Table 11.1, and for vulnerability and risk, see Appendix A.

3. The connections or dependencies between them need to be established for a specific threat, in this case, a coastal storm surge; express these dependencies in mathematical equations.

4. A system dynamics model is developed with the different assets as "systems" and by linking the different systems into an overall system; see Figures 11.2 and 11.3 as an example; the model uses the array structure for

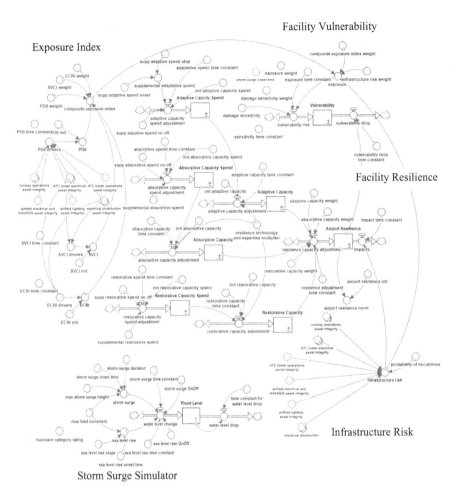

**FIGURE 11.2** Overall model structure with seven modules, with elements that generate system-wide measures for absorptive, adaptive, and restorative capacities and annual expenditures, vulnerability, resilience, and infrastructure risk. The model simulation is triggered by the storm surge simulator shown at the bottom left.

connecting the various assets across the different sectors (systems). For the basic storm surge module that generates the maximum storm surge height, onset time, and duration, see Figure 11.4. For the integrated infrastructure risk model, see Figure 11.5.

5. The simulation was carried out for a specific storm surge duration, and the impact on the assets was evaluated.
6. The results were plotted against time, for downtime/loss of AI, vulnerability, risk, and resilience; see Figures 11.6–11.9. The plots show the relative sensitivity of the different assets, the integrity values, and finally the overall airport resilience (see Figure 11.9).

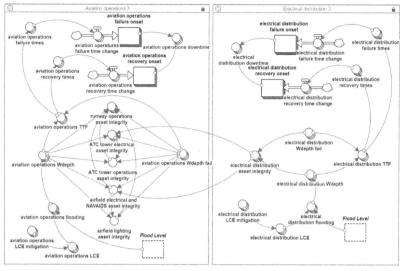

(Kautz and Mallick, 2017)

**FIGURE 11.3**    Aviation Operations sector (left sector panel), with 11 critical assets, and the Electrical Distribution sector (right sector panel), with four critical assets (electrical substations) for the airport infrastructure. The "stacked" icons for runway operations asset integrity denote six assets (runways), and those for the ATC tower electrical asset integrity denote two assets (substation and generator). Note that these two sectors are representative of several critical sectors that are present in a major airport. Using iThink software (Richmond et al., 2004), array structures can be built to map the relationships between multiple assets across multiple sectors.

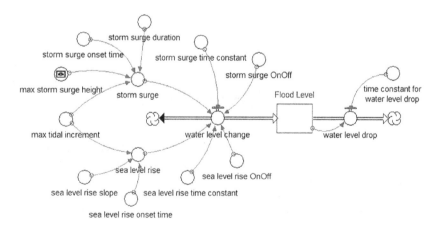

**FIGURE 11.4**    System dynamics model for capturing the impact of storm surge; the information flows through the red connectors and triggers the computation of simulation outputs for asset-specific water depth relative to the unique lowest critical elevation attributes, asset integrity measures, and asset downtime as well as the system-wide measures for absorptive, adaptive, and restorative capacities and spend; vulnerability; resilience; and infrastructure risk.

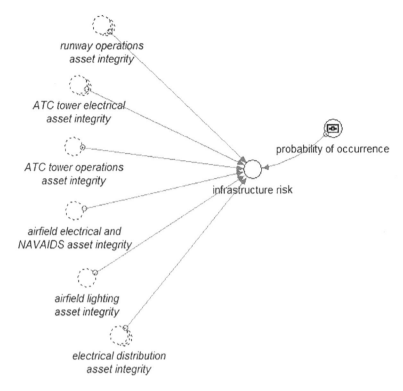

**FIGURE 11.5** Model segment combining asset-specific operation risks into an integrated system measure for infrastructure risk.

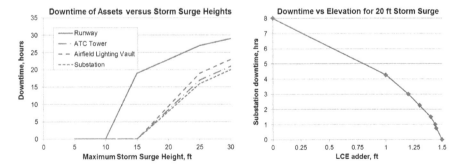

**FIGURE 11.6** Plot of maximum storm surge height versus downtime and lowest critical elevation versus substation downtime; note the sensitivity of the runway which becomes unoperational at a relatively lower storm surge, compared with the ATC tower, lighting, or substation, and hence creates an overall downtime impact on the entire airport infrastructure; the right-hand-side plot shows that for a 20-ft storm surge, with the addition of lowest critical of 1.5 ft (such as by adding flood wall) the downtime can be reduced to zero, that is, no impact.

Integrity

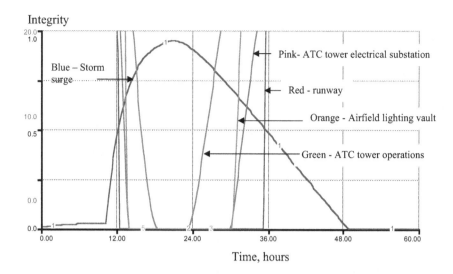

**FIGURE 11.7**    Plots of integrity over time for different airport assets for a 19-ft storm surge (full recovery).

Integrity

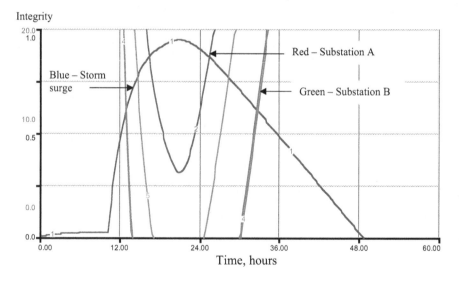

**FIGURE 11.8**    Plots of integrity over time for substations for a 19-ft storm surge.

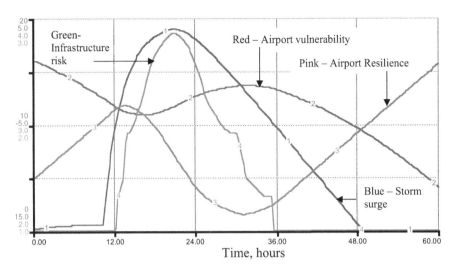

**FIGURE 11.9**   Vulnerability, resilience, and risk for the airport due to a 19-ft storm surge.

# REFERENCES

Mallick, R. B., Kautz, F., and Radzicki, M. (2018). Modeling Critical Infrastructure Resiliency: Application of System Dynamics to Multi-Asset Facility Response Under Coastal Storm Surge Threats. Paper presented at the Transportation Research Board 2018 Annual Meeting, Washington, DC.

Richmond, B., Peterson, S., Chichakly, K., Liu, W. and Wallis, J. (2004). *iThink Software*, Lebanon, NH: Isee Systems Inc.

Vugrin, E. D., Warren, D. E., Ehlen, M. A., and Camphouse, R. C. (2010). A framework for assessing the resilience of infrastructure and economic systems. In *Sustainable and Resilient Critical Infrastructure Systems* (K. Gopalakrishnan and S. Peeta, eds.), Berlin, Heidelberg: Springer, pp. 77–116.

# 12 System Dynamics – Path Forward

The realm of pavement engineering has crossed from physical sciences to social sciences, and hence we need a methodology (with appropriate tools) to consider both in solving complex problems. System dynamics is critical for modeling problems in pavement engineering because pavements involve society, economics, natural resources, and engineering – all of which need to be considered in the appropriate way for specific problems to get the right answers to our questions. As we have seen in the preceding chapters, system dynamics offers us the flexibility to cut across disciplines using logical expressions and mathematical connectors.

The ability to link different disciplines and evaluate the cascading impacts of different parts of the system makes system dynamics particularly relevant for evaluation of vulnerability and resilience of system and system of systems also. Today, when pavements are desirable to be designed, constructed, and maintained in a sustainable manner and when resilience, particularly against extreme weather events and climate change, is becoming extremely important for both the public and departments of transportation, system dynamics provides a most relevant methodology of **systems thinking, modeling, analyzing, simulating, and presenting results for the adoption of proper policies**.

System dynamics is based on principles of system thinking, logic, mathematics, physical, and social sciences (Forrester, 1961, 1969; Sterman, 2000; Morecroft, 2007). The steps that are inherent in system dynamics, that is defining the problem, developing a dynamic hypothesis modeling simulating, and presenting the results, bring out the critical factors over an extended period of time that are crucial for the development of good policies that can ultimately result in sustainable and resilient pavement systems. These features are important when working with people from different disciplines such as engineering, management, social science, and environmental science, and when the topic has interactions with users and maintainers of the asset (pavements), and when there is a need to inform people from other disciplines about the results of interactions in a complex problem. System dynamics helps us to engineer the systems properly, such that, with proper policies, the assets can be maintained properly over their life cycle (e.g., design, construction, maintenance, rehabilitation). In complex problems, where the objectives are often to satisfy conflicting criteria (more roads or less use of natural resources?), the modeling can evolve with system dynamics and then progress to more focused modeling methods – system dynamics helps us to first identify the most critical components and drivers of a system in a modeled problem. System dynamics can help evaluate risk, understand operation, and evaluate systems' impact and consequences on the society over the full life cycle. It captures the complex social interactions that occur during the functioning of a complex system and exchanges data between different disciplines.

DOI: 10.1201/9781003345596-12                    **119**

It is interesting to note here that system dynamics is one of the six modeling techniques that are regularly used by the National Aeronautics and Space Administration (NASA).

In this chapter, some of the most critical points about system dynamics, which show its relevance, uniqueness, and continuous improvement, are presented. While system dynamics is not certainly the only methodology to model complex problems in pavement engineering, this chapter argues that it is one of the most needed ones in education, research, and practice.

## MATHEMATICS TO MODEL

Regarding setting up a model, one of the relevant points to remember is that we can set up a model on the basis of a mathematical expression. For example, from a second-order differential equation, we can set up a model with two stocks and two flows as follows (*Radzicki, undated*).

$$a\frac{d^2Y}{dt^2} + b\frac{dy}{dt} + cy + d = 0$$

Consider $X = \dfrac{dy}{dt}$

Hence, $a\dfrac{dX}{dt} + bX + cy + d = 0$

$$a\frac{dX}{dt} = -bX - cy - d$$

$$\frac{dX}{dt} = \frac{(-bX - cy - d)}{a}$$

$X$ and $Y$ are considered to be stocks, and $\dfrac{dX}{dt}$ and $\dfrac{dY}{dt}$ are considered to be the corresponding flows.

In terms of abstract mathematics, which sets the basis for systems thinking and system dynamics, the reader is advised to refer to category theory (Eilenberg and MacLane, 1945), which explains how different algebraic (original definition) structures are related to each other, which can be expressed through domain/co-domain, identity, associativity, and composition (with mathematical operations) – in a way that is much richer in structure than set theory. Category theory has evolved significantly over time with the help of logicians, mathematicians, and mathematical physicists (for example, higher-dimensional categories have been introduced, Lawvere, F. W. 1966,

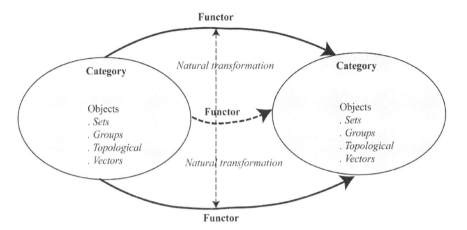

**FIGURE 12.1** Schematic representation of category and functor in category theory.

*The Category of Categories as a Foundation for Mathematics, Proceedings of the Conference on Categorical Algebra*, La Jolla, New York: Springer-Verlag, 1–21). A category is made up of groups with structure-preserving maps between them, and functors are used as maps that preserve the structure between the categories and, hence, the commutativity. There can be multiple functors between any two categories (Figure 12.1). Examples of functors include homology and homotopy. In system dynamics application, category theory helps us to identify the physical elements (objects/stocks) and the functors/maps (verbs or the functional relationships/flows) and, hence, create a logically simplified model structure (defined by composition). It further helps in the identification of isomorphism, to help improve efficiency in big data processing. Systems are, in essence, modeled with category theory structure in system dynamics.

## IDEAL STEPS FOR CREATING AND USING A SYSTEM DYNAMICS MODEL

The steps in the construction of system dynamics model (Figure 12.2) should ideally consist of defining the problem, developing a hypothesis, selecting the components and their relationships, creating the model, simulating and testing it, trying conclusions, and evaluating the robustness and the reference behavior of the model. There is a need for iterations between these steps. In addition, it is required to graph the behavior of the important variables over time.

The three fundamental questions that are needed for the development of hypothesis relate to (a) the components of the model; (b) details of the components of the model; and (c) relationships between the components of the model. The robustness of the model should be tested by checking the limits of the conditions under which it is valid. It can be tested by applying changes from the steady state to some extreme conditions.

Models should also be tested for reference behavior in which the model is able to replicate the historical data from experimentation or experience. It is advisable to check the reference behavior by normalizing the important variables and evaluating

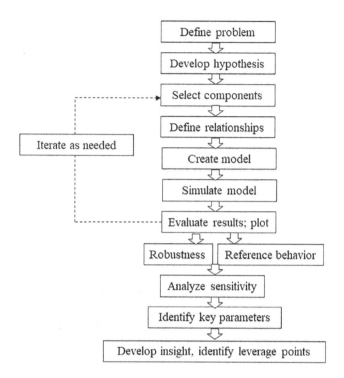

**FIGURE 12.2**    Suggested steps in system dynamics modeling.

their relative growth or decrease as a result of the simulation of the model compared with the historical data.

It is in the hypothesis development phase that feedback loop(s) is/are to be introduced, as appropriate. It is important to first develop a chain of stocks and flows and identify the key ones and then the important feedback loops. The units of the different stocks and flows should be in the proper order so that dimensional balance is maintained. Next, the use numerical values that would set the model in a steady state should be used. In the first place, these numbers should be making sense relative to each other. To initialize models in a steady-state condition, we need some values of parameters. We may start with those for which we are more confident in this process; we may use the values that we are comfortable with or confident with and then try to solve for the ones that we are not confident with using algebraic equation data used in the model.

The model can be simulated using different levels of variables to check for errors in the model. In the next step, results should be used for plotting graphs to check the results with the hypothesis. Next, the model should be simulated using extreme values to check the robustness of the model. Remember while checking for the reference behavior, it is important to check for qualitative similarity first. Finally, carry out a sensitivity study in which check again the qualitative or behavioral pattern first and then focus on the quantitative pattern; identify and analyze the most sensitive parameters, by disaggregating them and reducing them into simpler parameters.

## SYSTEMS SCIENCE, SYSTEMS ENGINEERING, AND SYSTEM DYNAMICS

Today, we face multiple challenges from resource crunch to environmental deterioration and climate change. Civil engineers have a vital role to play not only to face these challenges but also help in maintaining and flourishing the natural environment. To accomplish these tasks in the most efficient way, system dynamics enables us to analyze problems from a systems perspective – to make decisions regarding the best way of operation of the system.

Both the built and the natural system are responsibilities of civil engineers – how do we evaluate the performance of this existing joint system and redesign them for better results? For the different components in the lives of projects, such as design, build, operation, and maintenance, how do we integrate the decisions? System dynamics helps us to model the problems from a holistic systems perspective and identify the **leverage points** – where a small change can produce a significant change in the system. Furthermore, once we know how the components of a system affect each other, and the leverage points, by simulation experiments with the model, system dynamics can help us to know what should be the **order** of implementation/execution of policies that can use the leverage points. In addition, models help us to see opportunities for improvements of the system and/or savings that we would otherwise miss. This would lead to integrated planning that results in improvements in multiple aspects of the system. In subsequent iterations of the model, more details and numeric data can be added to have the results with more resolutions. In summary, pavements, society, and environment are an interconnected system, and the interactions need to be understood, such that better integrated planning and policies could be implemented to improve the system – in ways that are not possible to determine or predict without system dynamics modeling.

Systems science is involved with the exploration of systems research (simple modeling but difficult explanation) and complexity science (difficult to model but simple to explain), which fall under the realm of organized complexity, as opposed to disorganized complexity (classical statistics) and deterministic (classical analytical sciences). Systems engineering bridges the gap between these two aspects of systems science, with the help of both statistics and analytical approach, and engineering principles. The system concept, together with specific mechanisms such as feedback, brings us to the domain of system dynamics. Hence, systems science is a component of system dynamics. Similarly, the systems concept together with engineering concepts such as life cycle, gets us to systems engineering, whereas concepts such as perspectives bring us to systems thinking. Hence, system dynamics, systems engineering, and systems thinking are all tied together with systems science at the core. In a way, system dynamics also helps in systems engineering by enabling a holistic view of the problem, by bringing in social and policy perspectives, and systems thinking. It helps us cross the disciplinary boundary to enable us to move from defining the complex problem to a desirable solution in systems engineering (Figure 12.3).

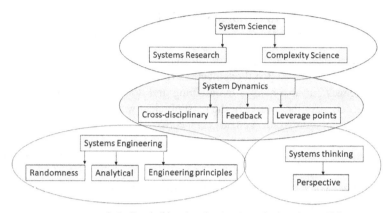

System Dynamics links system science to systems engineering and systems thinking
through its cross-disciplinary ability, and feedback and leverage point identification features

**FIGURE 12.3**   Importance of system dynamics.

## LIMITS TO GROWTH THEORY

Long before the establishment of the formal definition of sustainability by the Brundtland Commission (1987), system dynamics was used to derive a model on limits to economic growth (Meadows et al., 1972), whose relevance is as vital to pavement engineering as it is for any realm of economic and social development today. The model, which has been updated historically many times by different researches, clearly shows that resource consumption is exponential in nature and, left unchecked or unmodified, then at some point, the users will be forced to stop using the resources simply because they run out (see Figure 12.4; see Saeed, 1985 for system dynamics model). The original model shows that with respect to food/people/industry, when resources run out, there will be a catastrophic decline in population because of the want of food/agriculture/jobs because of loss of industry due to exhausting resources. The limits to growth model have been evaluated by many system dynamics researchers from different points of view, and one of the most rational ways that has been suggested, in which this catastrophe can be avoided is by matching the rate of use of resources with the rate of regeneration of resources and by increasing the efficiency of the use of the resources. This recommendation is particularly relevant for pavements in two aspects: (a) consider the timing of maintenance and rehabilitation of existing pavements in such a way that adequate recycled materials are available at that point and that no amount of virgin materials are required and (b) improve materials and structural design so that the efficiency of using construction materials, energy, and money is continually improved.

## USE OF SYSTEM DYNAMICS, SOURCE OF INFORMATION, AND SUGGESTED READINGS

System dynamics is very much a growing field. Numerous entities around the world are using, incorporating in, and merging with other techniques for tackling

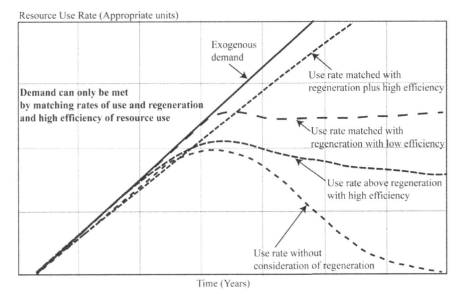

Resource Use Rate (Appropriate units)

Exogenous demand

Use rate matched with regeneration plus high efficiency

Demand can only be met by matching rates of use and regeneration and high efficiency of resource use

Use rate matched with regeneration with low efficiency

Use rate above regeneration with high efficiency

Use rate without consideration of regeneration

Time (Years)

**FIGURE 12.4** Exogenous demand and impact of different policies of rates of resource use. (Redrawn from Saeed, 1985.)

some of the most critical problems of our lives today, such as ensuring water resources, climate change vulnerabilities, adaptation, and resilience. Other important applications include socio-political problems, socio-economic problems, business dynamics, crime – law and order, biomedical opportunities, and pandemic (COVID-19) problems and impacts. Some of the prominent groups using system dynamics include MIT System Dynamics group (World Climate group), International Council on Systems Engineering (strategydynamics.com), NASA, and the Center for Systems Engineering and Innovation at Imperial College London. An excellent source of information on this topic is the System Dynamics Society (https://systemdynamics.org/), which provides an annotated bibliography of publications, regular seminars, and the *System Dynamics* review journal, which presents latest articles from experts. Members of this society can also become members of more focused Special Interest Groups such as asset dynamics, business, energy, education and security, stability, and resilience. Suggested must readings include *Principles of Systems and Industrial Dynamics* by J. W. Forrester, *Business Dynamics: Systems Thinking and Modeling for a Complex World* by J. D. Sterman, *Strategic Modelling and Business Dynamics: A Feedback Systems Approach* by J. D W. Morecroft, *Limits to Growth* by Donella H. Meadows, and *Towards Sustainable Development: Essays on System Analysis of National Policy* by Dr. Khalid Saeed. Readers are also advised to study the C-Roads Climate Change Policy Simulator (https://www.climateinteractive.org/c-roads/), an excellent example of system dynamics application for exploring the impacts of climate change–related strategies.

## REFERENCES

Eilenberg, S. and Mac Lane, S. (1945). General theory of natural equivalences. *Transactions of the American Mathematical Society*, 58, 231–294.

Forrester, J. W. (1961). *Industrial Dynamics*. Cambridge, MA: M.I.T. Press.

Forrester, J. W. (1969). *Principles of Systems*. Cambridge, MA: M.I.T. Press.

Meadows, D. H., Meadows, D. L., Randers, J., and Behrens, W. (1972). *The Limits to Growth; A Report for the Club of Rome's Project on the Predicament of Mankind*. New York: Universe Books.

Morecroft, J. D. W. (2007). *Strategic Modelling and Business Dynamics: A Feedback Systems Approach*. Chichester, England; Hoboken, NJ: John Wiley & Sons, Inc.

Saeed, K. (1985). An attempt to determine criteria for sensible rates of use of material resources. *Technological Forecasting and Social Change*, 28, 311–325.

Sterman, J. (2000). *Business Dynamics: Systems Thinking and Modeling for a Complex World*. Boston: Irwin/McGraw-Hill.

# Appendix A
## Equations for Model in Chapter 11

### TOP-LEVEL MODEL

Absorptive_Capacity(t) = Absorptive_Capacity(t - dt) + (absorptive_capacity_adjustment) * dt
    INIT Absorptive_Capacity = 30
    INFLOWS:
        absorptive_capacity_adjustment = (Absorptive_Capacity_Spend*resilience_technology_and_expertise_multiplier - init_absorptive_capacity)/absorptive_capacity_time_constant
Absorptive_Capacity_Spend(t) = Absorptive_Capacity_Spend(t - dt) + (absorptive_capacity_spend_adjustment) * dt
    INIT Absorptive_Capacity_Spend = 10
    INFLOWS:
        absorptive_capacity_spend_adjustment = (init_absorptive_capacity_spend + supplemental_absorptive_spend*supp_absorptive_spend_on_off)/absorptive_spend_time_constant
Adaptive_Capacity(t) = Adaptive_Capacity(t - dt) + (adaptive_capacity_adjustment) * dt
    INIT Adaptive_Capacity = 50
    INFLOWS:
        adaptive_capacity_adjustment = (Adaptive_Capcity_Spend*resilience_technology_and_expertise_multiplier + init_adaptive_capacity)/adaptive_capacity_time_constant
Adaptive_Capcity_Spend(t) = Adaptive_Capcity_Spend(t - dt) + (adaptive_capacity_spend_adjustment) * dt
    INIT Adaptive_Capcity_Spend = 10
    INFLOWS:
        adaptive_capacity_spend_adjustment = (init_adaptive_capacity_spend + supplemental_adaptative_spend*supp_adaptive_spend_on_off)/adaptative_spend_time_constant

aviation_operations_failure_onset[runway_asset_failure_onset](t) = aviation_operations_failure_onset[runway_asset_failure_onset](t - dt) + (aviation_operatures_failure_time_change[runway_asset_failure_onset]) * dt
    INIT aviation_operations_failure_onset[runway_asset_failure_onset] = 0

aviation_operations_failure_onset[airfield_electrical_and_NAVAIDs_asset_failure_onset](t) = aviation_operations_failure_onset[airfield_electrical_and_NAVAIDs_asset_failure_onset]

(t - dt) + (aviation_operatures_failure_time_change[airfield_
electrical_and_NAVAIDs_asset_failure_onset]) * dt
    INIT aviation_operations_failure_onset[airfield_
electrical_and_NAVAIDs_asset_failure_onset] = 0

aviation_operations_failure_onset[ATC_tower_asset_failure_
onset](t) = aviation_operations_failure_onset[ATC_tower_asset_
failure_onset](t - dt) + (aviation_operatures_failure_
time_change[ATC_tower_asset_failure_onset]) * dt
    INIT aviation_operations_failure_onset[ATC_tower_asset_
failure_onset] = 0

aviation_operations_failure_onset[ATC_tower_substation_asset_
failure_onset](t) = aviation_operations_failure_onset[ATC_
tower_substation_asset_failure_onset](t - dt) + (aviation_
operatures_failure_time_change[ATC_tower_substation_asset_
failure_onset]) * dt
    INIT aviation_operations_failure_onset[ATC_tower_
substation_asset_failure_onset] = 0

aviation_operations_failure_onset[ATC_tower_MPA_generator_
asset_failure_onset](t) = aviation_operations_failure_
onset[ATC_tower_MPA_generator_asset_failure_onset](t - dt) +
(aviation_operatures_failure_time_change[ATC_tower_MPA_
generator_asset_failure_onset]) * dt
    INIT aviation_operations_failure_onset[ATC_tower_MPA_
generator_asset_failure_onset] = 0

aviation_operations_failure_onset[airfield_lighting_vault_
asset_failure_onset](t) = aviation_operations_failure_
onset[airfield_lighting_vault_asset_failure_onset](t - dt) +
(aviation_operatures_failure_time_change[airfield_lighting_
vault_asset_failure_onset]) * dt
    INIT aviation_operations_failure_onset[airfield_lighting_
vault_asset_failure_onset] = 0

    INFLOWS:
        aviation_operatures_failure_time_change[runway_asset_
failure_onset] = aviation_operations_failure_times[runway_TTF_
times]/DT
        aviation_operatures_failure_time_change[runway_asset_
failure_onset] = aviation_operations_failure_times[runway_TTF_
times]/DT
        aviation_operatures_failure_time_change[airfield_
electrical_and_NAVAIDs_asset_failure_onset] = aviation_
operations_failure_times[airfield_electrical_and_NAVAIDs_TTF_
times]/DT
        aviation_operatures_failure_time_change[ATC_tower_
asset_failure_onset] = aviation_operations_failure_times[ATC_
tower_TTF_times]/DT

```
        aviation_operatures_failure_time_change[ATC_tower_
substation_asset_failure_onset] = aviation_operations_failure_
times[ATC_tower_substation_TTF_times]/DT
        aviation_operatures_failure_time_change[ATC_tower_MPA_
generator_asset_failure_onset] = aviation_operations_failure_
times[ATC_tower_MPA_generator_TTF_times]/DT

        aviation_operatures_failure_time_change[airfield_
lighting_vault_asset_failure_onset] = aviation_operations_
failure_times[airfield_lighting_vault_TTF_times]/DT
aviation_operations_recovery_onset[runway_asset_recovery_
onset](t) = aviation_operations_recovery_onset[runway_asset_
recovery_onset](t - dt) + (aviation_operations_recovery_
time_change[runway_asset_recovery_onset]) * dt
        INIT aviation_operations_recovery_onset[runway_asset_
recovery_onset] = 0
aviation_operations_recovery_onset[airfield_electrical_and_
NAVAIDs_asset_recovery_onset](t) = aviation_operations_
recovery_onset[airfield_electrical_and_NAVAIDs_asset_recovery_
onset](t - dt) + (aviation_operations_recovery_time_change
[airfield_electrical_and_NAVAIDs_asset_recovery_onset]) * dt
        INIT aviation_operations_recovery_onset[airfield_
electrical_and_NAVAIDs_asset_recovery_onset] = 0
aviation_operations_recovery_onset[ATC_tower_asset_recovery_
onset](t) = aviation_operations_recovery_onset[ATC_tower_
asset_recovery_onset](t - dt) + (aviation_operations_recovery_
time_change[ATC_tower_asset_recovery_onset]) * dt
        INIT aviation_operations_recovery_onset[ATC_tower_asset_
recovery_onset] = 0
aviation_operations_recovery_onset[ATC_tower_substation_asset_
recovery_onset](t) = aviation_operations_recovery_onset[ATC_
tower_substation_asset_recovery_onset](t - dt) + (aviation_
operations_recovery_time_change[ATC_tower_substation_asset_
recovery_onset]) * dt
        INIT aviation_operations_recovery_onset[ATC_tower_
substation_asset_recovery_onset] = 0
aviation_operations_recovery_onset[ATC_tower_MPA_generator_
asset_recovery_onset](t) = aviation_operations_recovery_
onset[ATC_tower_MPA_generator_asset_recovery_onset](t - dt) +
(aviation_operations_recovery_time_change[ATC_tower_MPA_
generator_asset_recovery_onset]) * dt
        INIT aviation_operations_recovery_onset[ATC_tower_MPA_
generator_asset_recovery_onset] = 0
aviation_operations_recovery_onset[airfield_lighting_vault_
asset_recovery_onset](t) = aviation_operations_recovery_
onset[airfield_lighting_vault_asset_recovery_onset](t - dt) +
(aviation_operations_recovery_time_change[airfield_lighting_
vault_asset_recovery_onset]) * dt
        INIT aviation_operations_recovery_onset[airfield_lighting_
vault_asset_recovery_onset] = 0
```

INFLOWS:
    aviation_operations_recovery_time_change[runway_asset_
recovery_onset] = aviation_operations_recovery_times[runway_
TTR_recovery_times]/DT
    aviation_operations_recovery_time_change[runway_asset_
recovery_onset] = aviation_operations_recovery_times[runway_
TTR_recovery_times]/DT
    aviation_operations_recovery_time_change[airfield_
electrical_and_NAVAIDs_asset_recovery_onset] = aviation_
operations_recovery_times[airfield_electrical_and_NAVAIDs_TTR_
recovery_times]/DT
    aviation_operations_recovery_time_change[ATC_tower_
asset_recovery_onset] = aviation_operations_recovery_
times[ATC_tower_TTR_recovery_times]/DT
    aviation_operations_recovery_time_change[ATC_tower_
substation_asset_recovery_onset] = aviation_operations_
recovery_times[ATC_tower_substation_TTR_recovery_times]/DT
    aviation_operations_recovery_time_change[ATC_tower_
MPA_generator_asset_recovery_onset] = aviation_operations_
recovery_times[ATC_tower_MPA_generator_TTR_recovery_times]/DT
    aviation_operations_recovery_time_change[airfield_
lighting_vault_asset_recovery_onset] = aviation_operations_
recovery_times[airfield_lighting_vault_TTR_recovery_times]/DT
electrical_distribution_failure_onset[substation_failure_
onset](t) = electrical_distribution_failure_onset[substation_
failure_onset](t - dt) + (electrical_distribution_failure_
time_change[substation_failure_onset]) * dt
    INIT electrical_distribution_failure_onset[substation_
failure_onset] = 0

    INFLOWS:
        electrical_distribution_failure_time_change[substation_
failure_onset] = electrical_distribution_failure_
times[substation_TTF_times]/DT

electrical_distribution_recovery_onset[substation_recovery_
onset](t) = electrical_distribution_recovery_onset[substation_
recovery_onset](t - dt) +
(electrical_distribution_recovery_time_change[substation_
recovery_onset]) * dt
    INIT electrical_distribution_recovery_onset[substation_
recovery_onset] = 0

    INFLOWS:
        electrical_distribution_recovery_time_change[substation_
recovery_onset] = electrical_distribution_recovery_
times[substation_TTR_times]/DT

Flood_Level(t) = Flood_Level(t - dt) + (water_level_change -
water_level_drop) * dt
    INIT Flood_Level = 0.1

```
INFLOWS:
      water_level_change = (storm_surge*storm_surge_OnOff/
storm_surge_time_constant) + (sea_level_rise*sea_level_rise_
OnOff/sea_level_rise_time_constant)
   OUTFLOWS:
      water_level_drop = Flood_Level/time_constant_for_
water_level_drop
Resilience(t) = Resilience(t - dt) + (resilience_capacity_
adjustment - Impacts) * dt
   INIT Resilience = 2.5

INFLOWS:
      resilience_capacity_adjustment = (Adaptive_Capacity*
adaptive_capacity_weight + Absorptive_Capacity*absorptive_
capacity_weight + Restorative_Capacity*restorative_capacity_
weight)/resilience_adjustment_time_constant

   OUTFLOWS:
      Impacts = infrastructure_risk/impact_time_constant
Restorative_Capacity(t) = Restorative_Capacity(t - dt) +
(restorative_capacity_adjustment) * dt
   INIT Restorative_Capacity = 20

INFLOWS:
      restorative_capacity_adjustment = (Restorative_
Capacity_Spend*resilience_technology_and_expertise_multiplier -
init_restorative_capacity)/restorative_capacity_time_constant
Restorative_Capacity_Spend(t) = Restorative_Capacity_Spend(t -
dt) + (restorative_capacity_spend_adjustment) * dt
   INIT Restorative_Capacity_Spend = 20

INFLOWS:
      restorative_capacity_spend_adjustment = (init_
restorative_capacity_spend + supplemental_restorative_
spend*supp_restorative_spend_on_off)/restorative_spend_time_
constant

Vulnerability(t) = Vulnerability(t - dt) + (vulnerability_rise -
vulnerability_drop) * dt
   INIT Vulnerability = 1
   INFLOWS:
      vulnerability_rise = damage_sensitivity*exposure/
exposure_time_constant
   OUTFLOWS:
      vulnerability_drop = Resilience/vulnerability_drop_
time_constant
absorptive_capacity_time_constant = 8760/2
absorptive_capacity_weight = 0.2
absorptive_spend_time_constant = 8760
adaptative_spend_time_constant = 8760
adaptive_capacity_time_constant = 8760/2
adaptive_capacity_weight = 0.6
```

airfield_electrical_and_NAVAIDS_asset_integrity = IF
(aviation_operations_Wdepth[airfield_elec_and_NAVAIDs_DFL]>=
aviation_operations_Wdepth_fail[airfield_elec_and_NAVAIDs_DFL_
fail]) THEN 0 ELSE (1-(aviation_operations_Wdepth[airfield_
elec_and_NAVAIDs_DFL]/aviation_operations_Wdepth_fail
[airfield_elec_and_NAVAIDs_DFL_fail]))*electrical_
distribution_asset_integrity[substation_AI]

airfield_lighting_asset_integrity = IF (aviation_operations_
Wdepth[airfield_lighting_vault_DFL]>= aviation_operations_
Wdepth_fail[airfield_lighting_vault_DFL_fail]) THEN 0 ELSE
(1-(aviation_operations_Wdepth[airfield_lighting_vault_DFL]/
aviation_operations_Wdepth_fail[airfield_lighting_vault_DFL_
fail]))*airfield_electrical_and_NAVAIDS_asset_integrity

ATC_tower_electrical_asset_integrity[ATC_tower_substation_AI] =
IF (aviation_operations_Wdepth[ATC_tower_substation_DFL]>=
aviation_operations_Wdepth_fail[ATC_tower_substation_DFL_
fail]) THEN 0 ELSE (1-(aviation_operations_Wdepth[ATC_tower_
substation_DFL]/aviation_operations_Wdepth_fail
[ATC_tower_substation_DFL_fail]))*electrical_distribution_
asset_integrity[substation_AI]

ATC_tower_electrical_asset_integrity[ATC_tower_MPA_generator_
AI] = IF (aviation_operations_Wdepth[ATC_tower_MPA_generator_
DFL]>= aviation_operations_Wdepth_fail[ATC_tower_MPA_
generator_DFL_fail]) THEN 0 ELSE 1-(aviation_operations_
Wdepth[ATC_tower_MPA_generator_DFL]/aviation_operations_
Wdepth_fail[ATC_tower_MPA_generator_DFL_fail])

ATC_tower_operations_asset_integrity = IF (aviation_
operations_Wdepth[ATC_tower_DFL]>= aviation_operations_Wdepth_
fail[ATC_tower_DFL_fail]) THEN 0 ELSE (1-(aviation_
operations_Wdepth[ATC_tower_DFL]/aviation_operations_Wdepth_
fail[ATC_tower_DFL_fail]))*ATC_tower_electrical_asset_
integrity[ATC_tower_MPA_generator_AI]

aviation_operations_downtime[runway_asset_downtime] =
ENDVAL(aviation_operations_recovery_onset[runway_asset_
recovery_onset]) - ENDVAL(aviation_operations_failure_onset
[runway_asset_failure_onset])

aviation_operations_downtime[runway_asset_downtime] =
ENDVAL(aviation_operations_recovery_onset[runway_asset_
recovery_onset]) - ENDVAL(aviation_operations_failure_
onset[runway_asset_failure_onset])

aviation_operations_downtime[airfield_electrical_and_NAVAIDs_
asset_downtime] = ENDVAL(aviation_operations_recovery_
onset[airfield_electrical_and_NAVAIDs_asset_recovery_onset]) -
ENDVAL(aviation_operations_failure_onset[airfield_electrical_
and_NAVAIDs_asset_failure_onset])

```
aviation_operations_downtime[ATC_tower_asset_downtime] =
ENDVAL(aviation_operations_recovery_onset[ATC_tower_asset_
recovery_onset]) - ENDVAL(aviation_operations_failure_onset
[ATC_tower_asset_failure_onset])
aviation_operations_downtime[ATC_tower_substation_asset_
downtime] = ENDVAL(aviation_operations_recovery_onset[ATC_
tower_substation_asset_recovery_onset]) - ENDVAL(aviation_
operations_failure_onset[ATC_tower_substation_asset_failure_
onset])
aviation_operations_downtime[ATC_tower_MPA_generator_asset_
downtime] = ENDVAL(aviation_operations_recovery_onset[ATC_
tower_MPA_generator_asset_recovery_onset]) - ENDVAL
(aviation_operations_failure_onset[ATC_tower_MPA_generator_
asset_failure_onset])
aviation_operations_downtime[airfield_lighting_vault_asset_
downtime] = ENDVAL(aviation_operations_recovery_onset
[airfield_lighting_vault_asset_recovery_onset]) - ENDVAL
(aviation_operations_failure_onset[airfield_lighting_vault_
asset_failure_onset])
aviation_operations_failure_times[runway_TTF_times] = IF
(aviation_operations_TTF[runway_TTF]>0) AND (PREVIOUS
(aviation_operations_TTF[runway_TTF], 0)=0) THEN TIME ELSE 0
aviation_operations_failure_times[runway_TTF_times] = IF
(aviation_operations_TTF[runway_TTF]>0) AND (PREVIOUS
(aviation_operations_TTF[runway_TTF], 0)=0) THEN TIME ELSE 0
aviation_operations_failure_times[airfield_electrical_and_
NAVAIDs_TTF_times] = IF (aviation_operations_TTF[airfield_
electrical_and_NAVAIDs_TTF]>0) AND (PREVIOUS(aviation_
operations_TTF[airfield_electrical_and_NAVAIDs_TTF], 0)=0)
THEN TIME ELSE 0
aviation_operations_failure_times[ATC_tower_TTF_times] = IF
(aviation_operations_TTF[ATC_tower_TTF]>0) AND (PREVIOUS
(aviation_operations_TTF[ATC_tower_TTF], 0)=0) THEN TIME ELSE 0
aviation_operations_failure_times[ATC_tower_substation_TTF_
times] = IF (aviation_operations_TTF[ATC_tower_substation_
TTF]>0) AND (PREVIOUS(aviation_operations_TTF[ATC_tower_
substation_TTF], 0)=0) THEN TIME ELSE 0
aviation_operations_failure_times[ATC_tower_MPA_generator_TTF_
times] = IF (aviation_operations_TTF[ATC_tower_MPA_generator_
TTF]>0) AND (PREVIOUS(aviation_operations_TTF[ATC_tower_
MPA_generator_TTF], 0)=0) THEN TIME ELSE 0
aviation_operations_failure_times[airfield_lighting_vault_TTF_
times] = IF (aviation_operations_TTF[airfield_lighting_vault_
TTF]>0) AND (PREVIOUS(aviation_operations_TTF[airfield_
lighting_vault_TTF], 0)=0) THEN TIME ELSE 0
aviation_operations_flooding[runway_inundation] =
Flood_Level-aviation_operations_LCE[runway_LCE]
aviation_operations_flooding[runway_inundation] =
Flood_Level-aviation_operations_LCE[runway_LCE]
```

```
aviation_operations_flooding[airfield_elec_and_NAVAIDs_
inundation] = Flood_Level-aviation_operations_LCE[airfield_
elec_and_NAVAIDs_LCE]


aviation_operations_flooding[ATC_tower_inundation] =
Flood_Level-aviation_operations_LCE[ATC_tower_LCE]
aviation_operations_flooding[ATC_tower_substation_inundation] =
Flood_Level-aviation_operations_LCE[ATC_tower_substation_LCE]
aviation_operations_flooding[ATC_tower_MPA_generator_
inundation] = Flood_Level-aviation_operations_LCE[ATC_
tower_MPA_generator_LCE]
aviation_operations_flooding[airfield_lighting_vault_
inundation] = Flood_Level-aviation_operations_LCE[airfield_
lighting_vault_LCE]


aviation_operations_LCE[runway_LCE] = 10.0 + aviation_
operations_LCE_mitigation[runway_LCE_mitigation]
aviation_operations_LCE[runway_LCE] = 10.0 + aviation_
operations_LCE_mitigation[runway_LCE_mitigation]


aviation_operations_LCE[airfield_elec_and_NAVAIDs_LCE] = 10.0 +
aviation_operations_LCE_mitigation[airfield_elec_and_NAVAIDs_
LCE_mitigation]
aviation_operations_LCE[ATC_tower_LCE] = 11.37 +
aviation_operations_LCE_mitigation[ATC_tower_LCE_mitigation]
aviation_operations_LCE[ATC_tower_substation_LCE] = 12.45 +
aviation_operations_LCE_mitigation[ATC_tower_substation_LCE_
mitigation]
aviation_operations_LCE[ATC_tower_MPA_generator_LCE] = 13.3 +
aviation_operations_LCE_mitigation[ATC_tower_MPA_generator_
LCE_mitigation]
aviation_operations_LCE[airfield_lighting_vault_LCE] = 14.61 +
aviation_operations_LCE_mitigation[airfield_lighting_vault_
LCE_mitigation]
aviation_operations_LCE_mitigation[runway_LCE_mitigation] = 0
aviation_operations_LCE_mitigation[runway_LCE_mitigation] = 0
aviation_operations_LCE_mitigation[airfield_elec_and_NAVAIDs_
LCE_mitigation] = 3.0
aviation_operations_LCE_mitigation[ATC_tower_LCE_mitigation] = 4.0
aviation_operations_LCE_mitigation[ATC_tower_substation_LCE_
mitigation] = 4.0
aviation_operations_LCE_mitigation[ATC_tower_MPA_generator_
LCE_mitigation] = 4.0
aviation_operations_LCE_mitigation[airfield_lighting_vault_
LCE_mitigation] = 0


aviation_operations_recovery_times[runway_TTR_recovery_times] =
IF (aviation_operations_TTF[runway_TTF]=0) AND (PREVIOUS
(aviation_operations_TTF[runway_TTF], 0)>0) THEN TIME ELSE 0
```

aviation_operations_recovery_times[runway_TTR_recovery_times] =
IF (aviation_operations_TTF[runway_TTF]=0) AND (PREVIOUS
(aviation_operations_TTF[runway_TTF], 0)>0) THEN TIME ELSE 0

aviation_operations_recovery_times[airfield_electrical_and_
NAVAIDs_TTR_recovery_times] = IF (aviation_operations_
TTF[airfield_electrical_and_NAVAIDs_TTF]=0) AND (PREVIOUS
(aviation_operations_TTF[airfield_electrical_and_NAVAIDs_TTF],
0)>0) THEN TIME ELSE 0
aviation_operations_recovery_times[ATC_tower_TTR_recovery_
times] = IF (aviation_operations_TTF[ATC_tower_TTF]=0) AND
(PREVIOUS(aviation_operations_TTF[ATC_tower_TTF], 0)>0) THEN
TIME ELSE 0
aviation_operations_recovery_times[ATC_tower_substation_TTR_
recovery_times] = IF (aviation_operations_TTF[ATC_tower_
substation_TTF]=0) AND (PREVIOUS(aviation_operations_TTF
[ATC_tower_substation_TTF], 0)>0) THEN TIME ELSE 0
aviation_operations_recovery_times[ATC_tower_MPA_generator_
TTR_recovery_times] = IF (aviation_operations_TTF[ATC_tower_
MPA_generator_TTF]=0) AND (PREVIOUS(aviation_operations_TTF
[ATC_tower_MPA_generator_TTF], 0)>0) THEN TIME ELSE 0
aviation_operations_recovery_times[airfield_lighting_vault_
TTR_recovery_times] = IF (aviation_operations_TTF[airfield_
lighting_vault_TTF]=0) AND (PREVIOUS(aviation_operations_TTF
[airfield_lighting_vault_TTF], 0)>0) THEN TIME ELSE 0

aviation_operations_TTF[runway_TTF] = IF (aviation_operations_
Wdepth[runway_DFL]>= aviation_operations_Wdepth_fail[runway_
DFL_fail]) THEN TIME ELSE 0
aviation_operations_TTF[runway_TTF] = IF (aviation_operations_
Wdepth[runway_DFL]>= aviation_operations_Wdepth_fail[runway_
DFL_fail]) THEN TIME ELSE 0

aviation_operations_TTF[airfield_electrical_and_NAVAIDs_TTF] =
IF (aviation_operations_Wdepth[airfield_elec_and_NAVAIDs_
DFL]>= aviation_operations_Wdepth_fail[airfield_elec_and_
NAVAIDs_DFL_fail]) THEN TIME ELSE 0
aviation_operations_TTF[ATC_tower_TTF] = IF (aviation_
operations_Wdepth[ATC_tower_DFL]>= aviation_operations_Wdepth_
fail[ATC_tower_DFL_fail]) THEN TIME ELSE 0
aviation_operations_TTF[ATC_tower_substation_TTF] = IF
(aviation_operations_Wdepth[ATC_tower_substation_DFL]>=
aviation_operations_Wdepth_fail[ATC_tower_substation_DFL_
fail]) THEN TIME ELSE 0
aviation_operations_TTF[ATC_tower_MPA_generator_TTF] = IF
(aviation_operations_Wdepth[ATC_tower_MPA_generator_DFL]>=
aviation_operations_Wdepth_fail[ATC_tower_MPA_generator_DFL_
fail]) THEN TIME ELSE 0
aviation_operations_TTF[airfield_lighting_vault_TTF] = IF
(aviation_operations_Wdepth[airfield_lighting_vault_DFL]>=

```
aviation_operations_Wdepth_fail[airfield_lighting_vault_DFL_
fail]) THEN TIME ELSE 0

aviation_operations_Wdepth[runway_DFL] = IF (aviation_
operations_flooding[runway_inundation] > 0) THEN aviation_
operations_flooding[runway_inundation] ELSE 0
aviation_operations_Wdepth[runway_DFL] = IF (aviation_
operations_flooding[runway_inundation] > 0) THEN aviation_
operations_flooding[runway_inundation] ELSE 0

aviation_operations_Wdepth[airfield_elec_and_NAVAIDs_DFL] = IF
(aviation_operations_flooding[airfield_elec_and_NAVAIDs_
inundation] > 0) THEN aviation_operations_flooding[airfield_
elec_and_NAVAIDs_inundation] ELSE 0
aviation_operations_Wdepth[ATC_tower_DFL] = IF (aviation_
operations_flooding[ATC_tower_inundation] > 0) THEN aviation_
operations_flooding[ATC_tower_inundation] ELSE 0
aviation_operations_Wdepth[ATC_tower_substation_DFL] = IF
(aviation_operations_flooding[ATC_tower_substation_inundation]
> 0) THEN aviation_operations_flooding[ATC_tower_substation_
inundation] ELSE 0
aviation_operations_Wdepth[ATC_tower_MPA_generator_DFL] = IF
(aviation_operations_flooding[ATC_tower_MPA_generator_
inundation] > 0) THEN aviation_operations_flooding[ATC_tower_
MPA_generator_inundation] ELSE 0
aviation_operations_Wdepth[airfield_lighting_vault_DFL] = IF
(aviation_operations_flooding[airfield_lighting_vault_
inundation] > 0) THEN aviation_operations_flooding[airfield_
lighting_vault_inundation] ELSE 0

aviation_operations_Wdepth_fail[runway_DFL_fail] = 0.2
aviation_operations_Wdepth_fail[runway_DFL_fail] = 0.2
aviation_operations_Wdepth_fail[airfield_elec_and_NAVAIDs_DFL_
fail] = 1.0
aviation_operations_Wdepth_fail[ATC_tower_DFL_fail] = 3.0
aviation_operations_Wdepth_fail[ATC_tower_substation_DFL_fail]
= 3.0
aviation_operations_Wdepth_fail[ATC_tower_MPA_generator_DFL_
fail] = 3.0
aviation_operations_Wdepth_fail[airfield_lighting_vault_DFL_
fail] = 2.0
composite_exposure_index = PDII*PDII_weight + SVCI*SVCI_weight
+ ECRI*ECRI_weight
composite_exposure_index_weight = 5
damage_sensitivity = 0.2
damage_sensitivity_weight = 0.5

ECRI = ECRI_init*ECRI_drivers

ECRI_drivers = 0.25 + 6/(2.0 + electrical_distribution_asset_
integrity[substation_AI] + ATC_tower_electrical_asset_
```

integrity[ATC_tower_substation_AI] + ATC_tower_electrical_
asset_integrity[ATC_tower_MPA_generator_AI])
ECRI_init = 0.6
ECRI_time_constant = 104
ECRI_weight = 0.10

electrical_distribution_asset_integrity[substation_AI] = IF
(electrical_distribution_Wdepth[substation_DFL]>= electrical_
distribution_Wdepth_fail[substation_DFL_fail]) THEN 0 ELSE
1-(electrical_distribution_Wdepth[substation_DFL]/
electrical_distribution_Wdepth_fail[substation_DFL_fail])

electrical_distribution_downtime[substation_downtime] =
ENDVAL(electrical_distribution_recovery_onset[substation_
recovery_onset]) - ENDVAL(electrical_distribution_failure_
onset[substation_failure_onset])

electrical_distribution_failure_times[substation_TTF_times] =
IF (electrical_distribution_TTF[substation_TTF]>0) AND
(PREVIOUS(electrical_distribution_TTF[substation_TTF], 0)=0)
THEN TIME ELSE 0

electrical_distribution_flooding[substation_inundation] =
Flood_Level-electrical_distribution_LCE[substation_LCE]

electrical_distribution_LCE[substation_LCE] = 11.15

electrical_distribution_recovery_times[substation_TTR_times] =
IF (electrical_distribution_TTF[substation_TTF]=0) AND
(PREVIOUS(electrical_distribution_TTF[substation_TTF], 0)>0)
THEN TIME ELSE 0
electrical_distribution_TTF[substation_TTF] = IF (electrical_
distribution_Wdepth[substation_DFL]>= electrical_distribution_
Wdepth_fail[substation_DFL_fail]) THEN TIME ELSE 0

electrical_distribution_Wdepth[substation_DFL] = IF
(electrical_distribution_flooding[substation_inundation]>0)
THEN electrical_distribution_flooding[substation_inundation]
ELSE 0

electrical_distribution_Wdepth_fail[substation_DFL_fail] = 3.0

exposure = composite_exposure_index*composite_exposure_index_
weight + infrastructure_risk*infrastructure_risk_weight
exposure_time_constant = 4
exposure_weight = 0.8
HAT_Highest_Astronomical_Tide = 7.73
hurricane_category_rating = 1
impact_time_constant = 12
infrastructure_risk = probability_of_occurrence*(0.25 + 15/
(5+ runway_operations_asset_integrity[runway_AI] + runway_

operations_asset_integrity[runway AI] +  ATC_tower_electrical_
asset_integrity[ATC_tower_substation_AI] + ATC_tower_
electrical_asset_integrity[ATC_tower_MPA_generator_AI] +
ATC_tower_operations_asset_integrity + airfield_electrical_
and_NAVAIDS_asset_integrity + airfield_lighting_asset_
integrity + electrical_distribution_asset_integrity
[substation_AI] ))
infrastructure_risk_weight = 0.5
init_absorptive_capacity = 1
init_absorptive_capacity_spend = 12
init_adaptive_capacity = 1
init_adaptive_capacity_spend = 10
init_restorative_capacity = 1.5
init_restorative_capacity_spend = 8

resilience_init = 1
resilience_norm = Resilience/resilience_init
max_storm_surge_height = 20
max_tidal_increment = 0.5
MHHW_Mean_Higher_High_Water = 5.58
MHW_Mean_High_Water = 5.14
MLW_Mean_Low_Water = -4.36
MSL_Mean_Sea_Level = 0.51
MTL_Mean_Tide_Level = 0.5
PDII = PDII_init*PDII_drivers
PDII_drivers = 0.25 + 15/(5+ runway_operations_asset_
integrity[runway_AI] +  ATC_tower_electrical_asset_
integrity[ATC_tower_substation_AI] + ATC_tower_electrical_
asset_integrity[ATC_tower_MPA_generator_AI] + ATC_tower_
operations_asset_integrity + airfield_electrical_and_NAVAIDS_
asset_integrity + airfield_lighting_asset_integrity +
electrical_distribution_asset_integrity[substation_AI] )
PDII_init = 1
PDII_time_constant = 104
PDII_weight = 0.80
probability_of_occurrence = 1
resilience_adjustment_time_constant = 8760/30
resilience_technology_and_expertise_multiplier = 1.05
restorative_capacity_time_constant = 8760/2
restorative_capacity_weight = 0.2
restorative_spend_time_constant = 8760

runway_operations_asset_integrity[runway_AI] = IF (aviation_
operations_Wdepth[runway_DFL]>= aviation_operations_Wdepth_
fail[runway_DFL_fail]) THEN 0 ELSE (1-(aviation_operations_
Wdepth[runway_DFL]/aviation_operations_Wdepth_fail
[runway_DFL_fail]))*airfield_electrical_and_NAVAIDS_asset_
integrity*airfield_lighting_asset_integrity*ATC_tower_
operations_asset_integrity

runway_operations_asset_integrity[runway_AI] = IF (aviation_
operations_Wdepth[runway_DFL]>= aviation_operations_Wdepth_
fail[runway_DFL_fail]) THEN 0 ELSE (1-(aviation_operations_
Wdepth[runway_DFL]/aviation_operations_Wdepth_fail
[runway_DFL_fail]))*airfield_electrical_and_NAVAIDS_asset_
integrity*airfield_lighting_asset_integrity*ATC_tower_
operations_asset_integrity

sea_level_rise = max_tidal_increment + RAMP(sea_level_rise_
slope, sea_level_rise_onset_time)
sea_level_rise_OnOff = 0
sea_level_rise_onset_time = 1.0
sea_level_rise_slope = 0.01
sea_level_rise_time_constant = 1000
sensitivity_time_constant = 12
storm_surge = max_tidal_increment + STEP(max_storm_surge_
height, storm_surge_onset_time) - RAMP (max_storm_surge_
height/(2.5*storm_surge_duration), storm_surge_onset_time +
storm_surge_duration)
storm_surge_duration = 10
storm_surge_OnOff = 1
storm_surge_onset_time = 10
storm_surge_time_constant = 3
supp_absorptive_spend_on_off = 0
supp_adaptive_spend_on_off = 0
supp_adaptive_spend_onset = 24
supp_adaptive_spend_step = -5
supp_restorative_spend_on_off = 0
supplemental_absorptive_spend = 0
supplemental_adaptative_spend = 0 + STEP(supp_adaptive_spend_
step, supp_adaptive_spend_onset)
supplemental_restorative_spend = 0
SVCI = SVCI_init*SVCI_drivers

SVCI_drivers = 0.25 + 6/(2.0 + electrical_distribution_asset_
integrity[substation_AI] +] + ATC_tower_electrical_asset_
integrity[ATC_tower_substation_AI] + ATC_tower_electrical_
asset_integrity[ATC_tower_MPA_generator_AI])
SVCI_init = 0.85
SVCI_time_constant = 104
SVCI_weight = 0.10
time_constant_for_water_level_drop = 3
vulnerability_drop_time_constant = 4

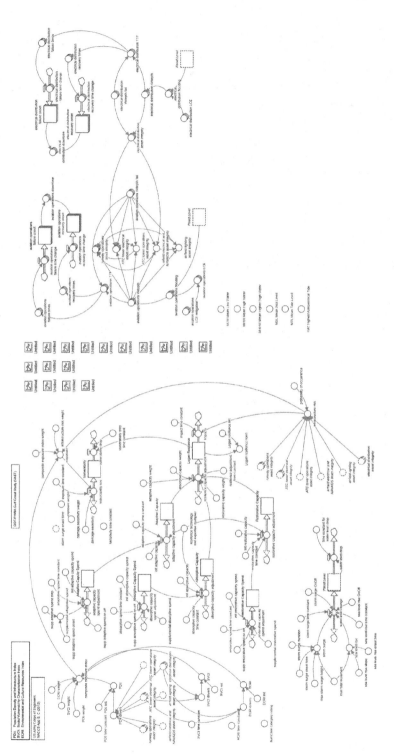

Printed in the United States
by Baker & Taylor Publisher Services